The Python Book
Python
数据分析

［英］罗布·马斯特罗多梅尼科　⊙著　宋廷强⊙译
（Rob Mastrodomenico）

清华大学出版社
北京

北京市版权局著作权合同登记号　图字：01-2022-6222

Title：The Python Book by Rob Mastrodomenico，ISBN：978-1-119-57331-9
Copyright © 2022 John Wiley and Sons Ltd
All Rights Reserved. This translation published under license. Authorized translation from the English language edition，Published by John Wiley & Sons．No part of this book may be reproduced in any form without the written permission of the original copyrights holder. Copies of this book sold without a Wiley sticker on the cover are unauthorized and illegal.

内 容 简 介

本书为想做数据处理但又没有编程经验的人提供了学习 Python 的入门指南。书中不仅涵盖了 Python 程序设计的相关内容，如变量、等式、列表、元组、字典、集合、while 和 for 循环，以及 if 语句等基础知识，还展示了如何利用 Python 进行数据分析、探索、清洗和整理等相关知识。

本书适用于从事数据统计、计算机、软件开发等相关人员，也适用于对相关领域感兴趣的读者。学习本书不需要编程基础，如果你经常与数据打交道并想要利用 Python 来提高工作效率，那么本书便是理想之选。

本书封面贴有 John Wiley & Sons 防伪标签，无标签者不得销售。
本书封面贴有清华大学出版社防伪标签，无标签者不得销售。
版权所有，侵权必究。举报：010-62782989，beiqinquan@tup.tsinghua.edu.cn。

图书在版编目（CIP）数据

Python 数据分析／（英）罗布·马斯特罗多梅尼科（Rob Mastrodomenico）著；宋廷强译. —北京：清华大学出版社，2024.4
　书名原文：The Python Book
　ISBN 978-7-302-65973-0

Ⅰ.①P… Ⅱ.①罗… ②宋… Ⅲ.①软件工具－程序设计 Ⅳ.①TP311.561

中国国家版本馆 CIP 数据核字（2024）第 068298 号

| 责任编辑：盛东亮　古　雪
| 封面设计：傅瑞学
| 责任校对：申晓焕
| 责任印制：刘　菲

出版发行：清华大学出版社
　　　　网　　　址：https://www.tup.com.cn，https://www.wqxuetang.com
　　　　地　　　址：北京清华大学学研大厦 A 座　　邮　编：100084
　　　　社 总 机：010-83470000　　　　　　　　　邮　购：010-62786544
　　　　投稿与读者服务：010-62776969，c-service@tup.tsinghua.edu.cn
　　　　质量反馈：010-62772015，zhiliang@tup.tsinghua.edu.cn
　　　　课件下载：https://www.tup.com.cn，010-83470236
印 装 者：三河市铭诚印务有限公司
经　　销：全国新华书店
开　　本：190mm×245mm　　印　张：15.75　　字　数：357 千字
版　　次：2024 年 4 月第 1 版　　　　　　　　印　次：2024 年 4 月第 1 次印刷
印　　数：1～2500
定　　价：79.00 元

产品编号：098268-01

译者序
PREFACE

　　本书是一部讲解 Python 基础知识及数据分析的入门教程。作者是一名数据科学家,使用过多种开发语言,编程经验丰富,对 Python 情有独钟,本书是作者使用 Python 进行数据分析的经验总结。书中对 Python 知识的介绍由浅入深,娓娓道来,非常适合没有编程经验的初学者入门学习,后面讲解的 Pandas 数据分析及数据可视化,采用多种公开数据集,引入了大量的开发实例,融入作者的开发实践,对于具有 Python 开发基础的程序员来说,也是一本非常不错的工具书。

　　译者对书中的所有代码都采用 Python 3.10.2 进行了验证,并对发现的问题都进行了纠正,简单问题直接进行了修改,部分更正进行了脚注说明,书中用到的数据集的含义及来源在脚注中尽量加以注释,一些英文关键词进行了中英文对照,便于读者阅读。有些运算依赖于运行环境,为了保证代码结果的正确性,部分采用了译者执行代码的结果。

　　书中的内容尽量做到通俗,符合汉语表达习惯,有些内容转换为表格显示会更加清晰。书中所有代码基于 Python 命令行模式、PyCharm2021、Spyder(Anaconda3)、Jupyter Notebook(Anaconda3)等进行开发验证,书中代码格式应用了 PyCharm、Spyder 等字体及配色设置,便于代码的可读性。

　　Python 解释器可以是命令行形式或者集成开发环境,为了让读者了解 Python 语言的特点,书中大部分代码都是以命令行的形式展示。命令行模式下单行代码回车可以直接执行,多行代码需要使用换行提示符"…",如果换行提示符后面没有代码,则表示多行代码的结束,此时回车便可以直接执行该多行代码。命令行模式便于查看代码的执行过程,如下所示:

```
>>> x = 2
>>> if x == 1:
...     x = x + 1
... else:
...     x = x - 1
...
>>> print("x = ",x)
x = 1
>>>
```

其中，">>>"是 Python 提示符，后面是 Python 语句。"…"是 Python 解释器的换行提示符，表示续写前面的 Python 语句。"x＝1"这一行没有以">>>"或"…"开头，表示是前面 Python 语句的执行结果。

如果示例代码语句中没有">>>"或"…"，表示这是一段 Python 代码程序，需要在 Python 集成开发环境下执行，通常下面都会紧跟该程序的执行结果及原理说明。

对于本书的出版译者首先要感谢清华大学出版社的编辑们，是他们的努力促成了本书的顺利翻译与出版，使读者能够通过本书学习 Python 开发技巧，同时也要感谢本书作者对于 Python 开发的总结和凝练。最后，感谢魏国政、盛兆康、刘儒一、宋家乐等对译稿进行核对与检查。

在本书的翻译过程中，虽力求忠于原著，但由于译者技术及翻译水平有限，书中难免存在疏漏，敬请读者批评指正。

译 者

2024 年 2 月于青岛

目 录
CONTENTS

第 1 章　绪论 1

第 2 章　准备开始 2

第 3 章　包和内置函数 6

第 4 章　数据类型 10

第 5 章　运算符 18

第 6 章　日期 24

第 7 章　列表 27

第 8 章　元组 36

第 9 章　字典 38

第 10 章　集合 44

第 11 章　循环与分支结构 54

第 12 章　字符串 63

第 13 章　正则表达式 69

第 14 章　文件操作 75

　　14.1　Excel 文件 79

- 14.2 JSON 文件 …… 81
- 14.3 XML 文件 …… 83

第 15 章 函数与类 …… 88

第 16 章 Pandas …… 99
- 16.1 NumPy 数组 …… 99
- 16.2 Series …… 102
- 16.3 DataFrame …… 107
- 16.4 concat()、merge()和 join()方法 …… 117
- 16.5 DataFrame 方法 …… 136
- 16.6 缺失值处理 …… 140
- 16.7 数据分组、聚合 …… 146
- 16.8 Pandas 文件操作 …… 154

第 17 章 数据可视化 …… 157
- 17.1 Pandas …… 157
- 17.2 Matplotlib …… 165
- 17.3 Seaborn …… 174

第 18 章 Python API …… 206

第 19 章 Python 网络爬虫 …… 219
- 19.1 HTML 简介 …… 219
- 19.2 网页抓取 …… 223

第 20 章 总结 …… 245

第 1 章 绪 论

  欢迎阅读《Python 数据分析》。在本书中,你将了解 Python 这门编程语言。这本书汇集了我十年来在 Python 教学和使用中的经验。作为一名数据科学家,我曾经使用过多种编程语言,Python 是最吸引我的。为什么要用 Python 呢?对我来说,我喜欢 Python,因为它开发速度很快,能够涵盖许多不同的应用程序,借助 Python 我可以完成几乎所有开发任务。然而,对于读者来说,Python 是一种很好的选择,因为它易于学习,上手很快,尤其是对于新手来说,可以充分享受语言学习过程中的成就感。本书不仅适用于新手,对于有经验的 Python 爱好者来说,也具有很高的参考价值。许多用户想要快速学习 Python,就必然要跳过许多基础知识。本书涵盖了 Python 的所有基础知识,可以为您强大的功能开发打下坚实的基础。Python 是一种简单优美的语言,本书在讲解内容时也坚持这一原则,尽量简单,不做复杂化描述,用简单的术语解释 Python 中的概念,并用实例加以验证。

  上面已经介绍了本书的目的,那么 Python 是什么?简单来说 Python 是一种编程语言,它的通用性意味着它可以做很多事。Python 应用广泛,本书专注于讲解 Python 数据处理的实际应用,除此之外,Python 也可以应用于人工智能、机器学习、Web 开发等。Python 是一种高级解释性语言,它的代码不需要被编译就可以执行。Python 最吸引人的地方是其语法简单,易于学习,代码编写方便。Python 语言语法简单,易于理解,语言风格清晰划一,利用缩进区分代码结构。Python 也是一种面向对象(object orientated)的语言,本书将详细地加以说明。不过需要指出的是,Python 既支持面向对象的编程,也支持面向过程的编程,用户可以根据需要自行选择。学习编程语言的最好方法就是实践,让我们安装好 Python,开始 Python 的学习之旅吧。

CHAPTER 2

第 2 章 准备开始

本书推荐读者安装 Python 的 Anaconda 发行版本,在 Anaconda 官方网站可根据需要下载 Windows、Mac 或 Linux 版本。安装 Anaconda 后,可以访问 Anaconda Navigator 页面,如图 2.1 所示。

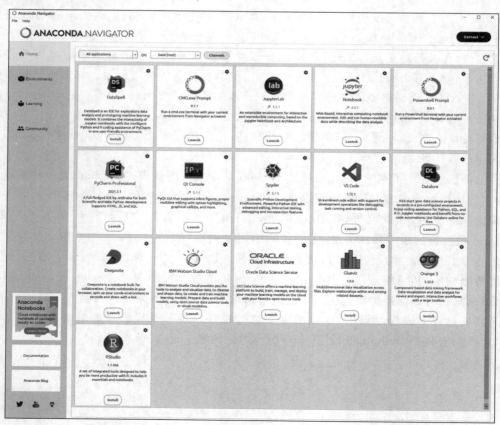

图 2.1　Anaconda Navigator 主页

默认安装工具包括以下内容：
- JupyterLab
- Notebook
- Qt Console
- Spyder

读者可以使用 Notebook 或者 Qt Console 实践本书中的示例。Notebook 是一个基于 Web 的交互式编辑器，如图 2.2 所示。

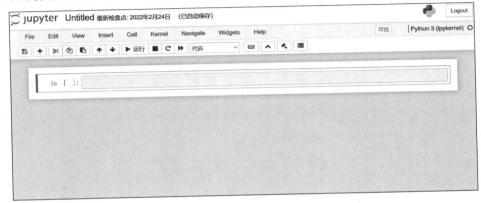

图 2.2 Notebook 界面

借助 Notebook 进行实践是一种常用且受欢迎的方式。读者可以键入代码，执行命令，然后查看结果。下面展示如何定义变量 x，键入 x 并使用运行按钮执行命令查看结果，如图 2.3 所示。

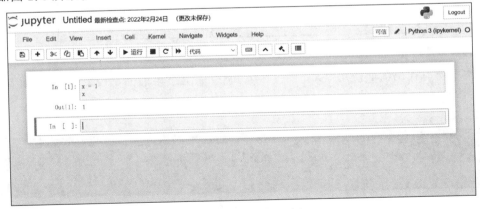

图 2.3 Notebook 示例

借助 Qt Console 进行实践是本书推荐的一种方式。Qt Console 通过控制台界面展示运行结果，运行界面如图 2.4 所示。

Qt Console 的使用与 Notebook 类似，图 2.5 中展示了使用 Qt Console 运行的相同示例。

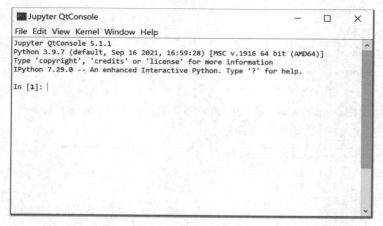

图 2.4　Qt Console 界面

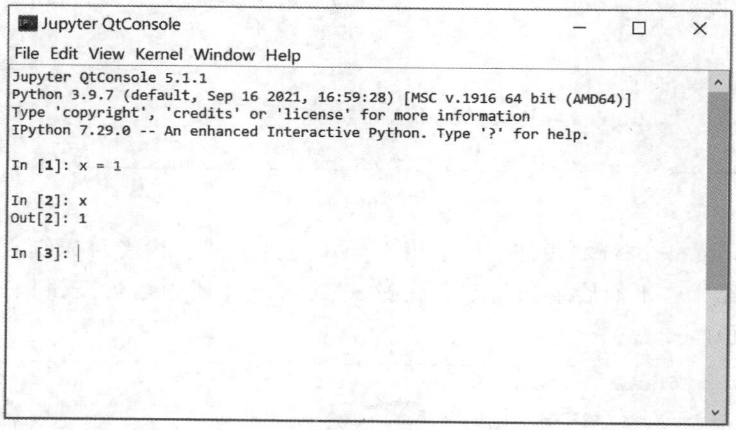

图 2.5　Qt Console 示例

本书所有代码前面使用"＞＞＞"进行标识，所有输出结果前面都没有该标识，如图 2.6 所示。

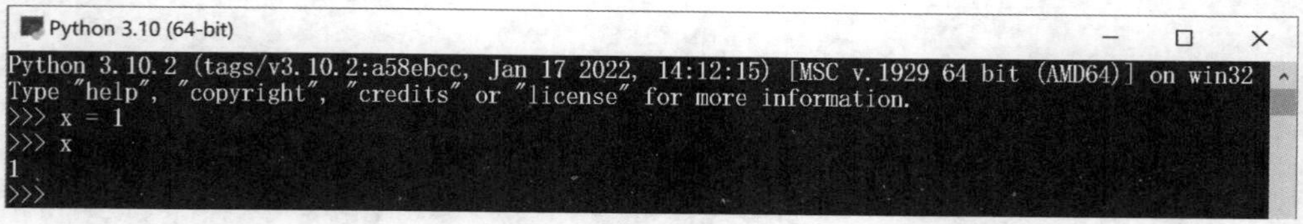

图 2.6　代码及输出

读者需要熟悉的另一个概念是如何使用终端（Linux 或 Mac 系统）或命令提示符窗口（Windows 系统）进行导航。使用时可以有多种途径，最简单的方法是利用终端搜索找到相关界面，在文件系统中可以使用 cd 命令切换目录，与双击打开文件夹查看里面有什么文件一样。与使用文件夹界面不同的是，默认情况下，用户无法看到某一目录中的内容,需要使用 ls 命令才能查看。ls 命令可以列出当前目录下的文件和子目录。图 2.7 用示例展示了如何打开目录,然后执行 Python 文件。

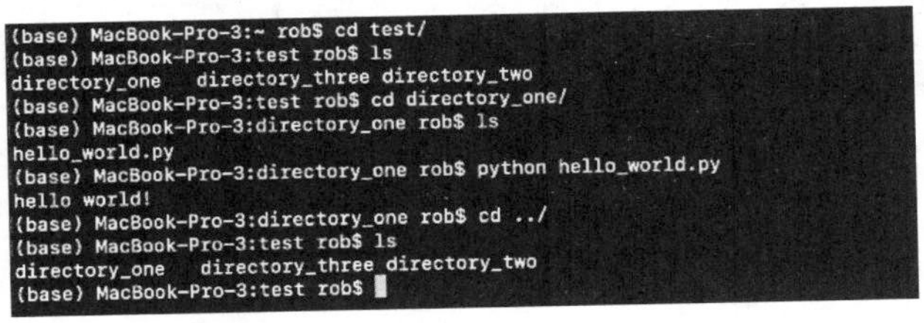

```
(base) MacBook-Pro-3:~ rob$ cd test/
(base) MacBook-Pro-3:test rob$ ls
directory_one    directory_three directory_two
(base) MacBook-Pro-3:test rob$ cd directory_one/
(base) MacBook-Pro-3:directory_one rob$ ls
hello_world.py
(base) MacBook-Pro-3:directory_one rob$ python hello_world.py
hello world!
(base) MacBook-Pro-3:directory_one rob$ cd ../
(base) MacBook-Pro-3:test rob$ ls
directory_one    directory_three directory_two
(base) MacBook-Pro-3:test rob$
```

图 2.7 命令行示例

除了 AnacondaNavigator 之外,我们还自动安装了 250 多个开源的数据科学包和机器学习包。借助 conda,用户可以将 7500 多个工具包轻松地安装到 Python 中。Anaconda 附带的相关操作系统的完整工具包列表和有关使用 conda installer 的详细信息,请访问官方网站,本书不作说明。本章最后简单介绍的概念是虚拟环境,用户可以在隔离的 Python 环境中开发并根据需要添加工具包。这是一种非常流行的开发方法,但由于本书面向初学者,所以本书将使用 Anaconda 安装包中所包含的工具包。

第 3 章 包和内置函数

前面已经提到了包(package)的概念,但并未加以详述,本节将详细说明包在 Python 配置中的作用。正如前面所述,Python 是一种面向对象的编程语言,在 Python 中万物皆对象,这一点在后面的编程实践中会深刻认识到。但是,有几个很重要的内置函数(built in functions)不是面向对象的,由于这几个函数在本书中会多次使用,需要先对其进行说明。这些内置函数的使用将会贯穿于本书内容,因此需要对其格外注意。下面给出这几个常用的内置函数。

- dir():该函数接收一个对象,并返回该对象属性和方法的列表。

```
>>> name = 'Rob'
>>> dir(name)
['__add__', '__class__', '__contains__', '__delattr__', '__dir__', '__doc__', '__eq__', '__format__', '__ge__', '__getattribute__', '__getitem__', '__getnewargs__', '__gt__', '__hash__', '__init__', '__init_subclass__', '__iter__', '__le__', '__len__', '__lt__', '__mod__', '__mul__', '__ne__', '__new__', '__reduce__', '__reduce_ex__', '__repr__', '__rmod__', '__rmul__', '__setattr__', '__sizeof__', '__str__', '__subclasshook__', 'capitalize', 'casefold', 'center', 'count', 'encode', 'endswith', 'expandtabs', 'find', 'format', 'format_map', 'index', 'isalnum', 'isalpha', 'isascii', 'isdecimal', 'isdigit', 'isidentifier', 'islower', 'isnumeric', 'isprintable', 'isspace', 'istitle', 'isupper', 'join', 'ljust', 'lower', 'lstrip', 'maketrans', 'partition', 'removeprefix', 'removesuffix', 'replace', 'rfind', 'rindex', 'rjust', 'rpartition', 'rsplit', 'rstrip', 'split', 'splitlines', 'startswith', 'strip', 'swapcase', 'title', 'translate', 'upper', 'zfill']
```

- float():将一个整型字符串转化为浮点数。①

```
>>> x = '1'
>>> float(x)
1.0
```

---

① 原著代码 float(1)修改为 float(x)。

- int()：将字符串对象转化为整数。[①]

```
>>> x = '1'
>>> int(x)
1
```

- len()：返回对象的长度。

```
>>> name = 'Rob'
>>> len(name)
3
>>> x = [1,2,3,4]
>>> len(x)
4
```

- list()：依据给定的参数创建列表。

```
>>> name = 'Rob'
>>> list(name)
['R', 'o', 'b']
```

- max()：从给定的参数中求取最大值。

```
>>> x = [1,2,3,4]
>>> max(x)
4
>>> name = ['r', 'o', 'b']
>>> max(name)
'r'
```

- min()：从给定的参数中求取最小值。

```
>>> x = [1,2,3,4]
>>> min(x)
1
>>> name = ['r', 'o', 'b']
>>> min(name)
'b'
```

- print()：将对象以文本的形式显示。

```
>>> x = [1,2,3,4]
>>> print(x)
[1, 2, 3, 4]
```

- round()：按照指定的精度对数值进行四舍五入。

[①] 原著代码 int(1)修改为 int(x)。

```
>>> y = 1.387668
>>> round(y,2)
1.39
```

- str()：将对象转化为字符串类型。

```
>>> y = 1.387668
>>> str(y)
'1.387668'
```

- type()：返回对象的类型。

```
>>> y = 1.387668
>>> type(y)
<class 'float'>
```

- abs()：返回传入的数值参数的绝对值。

```
>>> z = -0.657
>>> abs(z)
0.657
```

- help()：调用 Python 帮助系统给出帮助信息。

```
>>> help(list)
Help on class list in module builtins:

class list(object)
 |  list(iterable=(), /)
 |
 |  Built-in mutable sequence.
 |
 |  If no argument is given, the constructor creates a new empty list.
 |  The argument must be an iterable if specified.
 |
 |  Methods defined here:
 |
 |  __add__(self, value, /)
 |      Return self+value.
 |
 |  __contains__(self, key, /)
 |      Return key in self.
 |
 |  __delitem__(self, key, /)
 |      Delete self[key].
 |
 |  __eq__(self, value, /)
 |      Return self==value.
```

至此，如果对上述提到的 Python 相关的概念不是很熟悉也无碍，因为在接下来的章节中都会对这些内容进行详细介绍。

在给出这些内置函数的同时，Python 还提供了许多包。这些包执行一些特定的功能，并能够导入（import）到代码中。Python 提供了大量的包，当安装完系统之后就默认存在了。除此之外，还有大量的由第三方提供的包可以使用。在使用 Anaconda 发行版本时就安装了所有的默认包，包括在前面提到的一些包。本书会用到系统默认提供的包，也会用到第三方提供的包。为了说明这一点，接下来将介绍如何导入包。

下面要介绍的包是 datetime 包，该包是 Python 标准库的一部分，即该包来自 Python 本身，不是由第三方提供。只需要执行下面的语句就可以导入 datetime 包。

```
>>> import datetime
```

将 datetime 包导入之后，便可以访问 datetime 包中的所有内容。为了了解 datetime 包中有什么内容，可以借助前面讲到的内置函数 dir() 查看该对象的属性。

```
>>> import datetime
>>> dir(datetime)
['MAXYEAR', 'MINYEAR', '__all__', '__builtins__', '__cached__', '__doc__', '__file__', '__loader__', '__name__', '__package__', '__spec__', 'date', 'datetime', 'datetime_CAPI', 'sys', 'time', 'timedelta', 'timezone', 'tzinfo']
```

现在，如果想要查看这些属性的值具体是什么，则可以使用"."访问对象的属性。例如，要查看 MAXYEAR 和 MINYEAR 就可以进行如下操作：

```
>>> import datetime
>>> datetime.MAXYEAR
9999
>>> datetime.MINYEAR
1
```

导入包的时候，可以利用下面给出的语法格式，将包的指定部分导入。

```
>>> from datetime import date
```

这种导入的意思是从 datetime 包中仅导入了 date 属性，因而在后续的代码中也只能访问 datetime 包中的 date 属性。在实际代码编写中，仅导入包中需要的部分是一种好的习惯。这样，每当需要用到日期时就可以简单地直接调用 date，还可以在导入包的时候引入别名以简化代码量，如引入别名"d"代替"date"，示例如下：

```
>>> from datetime import date as d
```

以上是导入包的基本格式，在本书中会导入不同的包，也会给出如何利用类似的格式导入自己开发的代码。包与内置函数一样都是十分重要的概念，在本书中都会大量使用。

CHAPTER 4

第 4 章 数 据 类 型

接下来要介绍的概念是数据类型(Data Type)。本章将介绍一些常用的数据类型(整数、浮点数和字符串),并展示其进行基本运算(加、减、乘、除)时的特性。

1. 常用的数据类型

(1) 整数(integer):整数是不带小数点的数字。示例如下:

```
>>> 1
1
>>> 2
2
```

(2) 浮点数(float):浮点数是带小数点的数字。如果将上述整数写成浮点数的形式,则示例如下:

```
>>> 1.0
1.0
>>> 2.0
2.0
```

(3) 字符串(string):字符串是由单引号或双引号括起来的一些字符。同样,如果将上述浮点数写成字符串的形式,则示例如下:

```
>>> "2.0"
'2.0'
```

思考一下:假定我们知道怎样定义这些数据类型的变量,那么该如何检查变量的类型呢? 实际上,Python 提供了一个 type()方法用于确定变量的类型。因此,可以把之前示例中的数据重新赋值给一个变量,并调用 Python 的 type()方法查看其类型输出。

```
>>> x = 1
>>> type(x)
```

```
<class 'int'>
>>> y = 1.0
>>> type(y)
<class 'float'>
>>> z = "1.0"
>>> type(z)
<class 'str'>
```

2．基本运算

通过上面的学习，我们学会了如何定义变量，但关键是如何使用它们。常用的数学基本运算有加、减、乘、除，它们的操作符分别如下：

- ＋
- －
- ＊
- ／

下面介绍如何使用这 4 个操作符进行基本运算。

（1）"＋"运算。用"＋"连接两个整数表示数学上的加法。示例如下：

```
>>> x = 10
>>> y = 16
>>> x + y
26
```

同样地，如果将两个整数改成两个浮点数会得到类似的结果。示例如下：

```
>>> x = 10.0
>>> y = 16.0
>>> x + y
26.0
```

但是，如果参与加法的两个数一个为整数另一个为浮点数，会产生怎样的结果呢？示例如下：

```
>>> x = 10
>>> y = 16.0
>>> z = x + y
>>> z
26.0
>>> type(z)
<class 'float'>
```

从以上结果可以看到，浮点数与整数相加结果为浮点数，因此将整数转化成了浮点数。

如果对字符串进行加法运算会有什么结果呢？将上述示例中的 x 和 y 换成字符串，

执行加法运算，结果如下：

```
>>> x = "10"
>>> y = "16.0"
>>> z = x + y
>>> z
'1016.0'
```

由上可以看出，结果是把 x 和 y 两个字符串连接（拼接）起来了，这是字符串处理中常用的方法。

前面考虑了将整数与浮点数利用"＋"操作符运算的情况。如果利用"＋"操作符对字符串和整数进行运算，会出现什么情况呢？

```
>>> x = "10"
>>> y = 16
>>> z = x + y
Traceback(most recent call last):
    File "< stdin >", line 1, in < module >
TypeError: can only concatenate str(not "int") to str
```

此时出现了错误提示信息，指出不能将字符串与整数对象进行拼接。从本例可以看出，Python 试图利用"＋"操作符进行拼接操作，但由于两个操作数不全是字符串而不能完成该操作，并因此抛出错误信息。在 Python 中，不能将字符串与整数或者将字符串与浮点数混合运算，因此在后续章节中不考虑这种情况的运算。

（2）"－"运算。首先来看一下两个操作数都是整数的情况：

```
>>> x = 10
>>> y = 16
>>> z = x - y
>>> z
-6
```

正如所期望的一样，两个整数执行"－"运算的结果与数学减法相同。同样，如果对两个浮点数或者浮点数与整数执行"－"运算，其结果都等同于数学减法。

那么，对于字符串会怎样呢？能否对两个字符串执行"－"运算呢？示例如下：

```
>>> x = "10"
>>> y = "16"
>>> z = x - y
Traceback(most recent call last):
    File "< stdin >", line 1, in < module >
TypeError: unsupported operand type(s) for - : 'str' and 'str'
```

此时程序产生了另一种错误，但错误的原因是"－"运算不支持字符串。也就是说，Python 无法对两个字符串执行"－"运算；同样，Python 也无法对字符串执行"＊"和"/"

运算。因此，从本章所学内容来看，如果要进行字符串处理，只能利用"＋"进行字符串的拼接操作。

（3）"＊"运算。对于两个整数的"＊"运算，示例如下：

```
>>> x = 10
>>> y = 16
>>> x * y
160
```

乘法运算的原理很简单，对于两个浮点数同样适用。如果将浮点数与整数执行乘法运算，则有：

```
>>> x = 10
>>> y = 16.0
>>> x * y
160.0
```

可以看出，返回的乘法结果为浮点数。与加法和减法一样，都是将整数转换成了浮点数。

（4）"/"运算。下面介绍"/"运算怎样处理整数和浮点数的问题。当数据类型相同时，对两个整数进行"/"运算。示例如下：

```
>>> x = 10
>>> y = 16
>>> x / y
0.625
```

除此之外还有其他数据类型，例如复数。复数的定义示例如下：

```
>>> x = 3 + 5j
>>> x
(3+5j)
>>> y = 5j
>>> z = -5j
>>> z
(-0-5j)
```

借助下述方法，可以得到复数的实部和虚部。

```
>>> x.real
3.0
>>> x.imag
5.0
>>> y.real
0.0
>>> y.imag
```

```
5.0
>>> z.real
-0.0
>>> z.imag
-5.0
```

可以使用内置函数 complex() 构建一个复数,示例如下:

```
>>> a = 3
>>> b = 5
>>> c = complex(a, b)
>>> c
(3+5j)
>>> c.real
3.0
>>> c.imag
5.0
```

在运算方面,复数同样支持前面给出的标准运算符,示例如下:

```
>>> x = 3+5j
>>> x
(3+5j)
>>> y = 5j
>>> z = -5j
>>> z
(-0-5j)
>>> x + y
(3+10j)
>>> x - y
(3+0j)
>>> x / y
(1-0.6j)
>>> x * y
(-25+15j)
```

复数还可以与整数或浮点数进行加、减、乘、除运算,示例如下:

```
>>> x + 10.2
(13.2+5j)
>>> x - 10
(-7+5j)
>>> x - 10.2
(-7.199999999999999+5j)
>>> x * 10
(30+50j)
>>> x
(3+5j)
>>> x * 10.2
```

```
(30.599999999999998+51j)
>>> x * 10.2
(30.599999999999998+51j)
>>> x / 10
(0.3+0.5j)
>>> x / 10.2
(0.29411764705882354+0.4901960784313726j)
```

可以看出,当复数与整数或浮点数进行加、减运算时,结果只改变了复数的实部;而当复数与整数或浮点数进行乘、除运算时,复数的实部和虚部都将发生改变。

接下来介绍 Python 中的布尔值。布尔值可以使用 True 或 False 进行定义,示例如下:

```
>>> x = True
>>> x
True
>>> y = False
>>> y
False
```

借助内置函数 bool() 可以将整数或浮点数转换为布尔值,该函数将任何视为 0 或 0.0 的数值转换为 False,将其他值转换为 True。示例如下:

```
>>> x = bool(1)
>>> x
True
>>> y = bool(0.0)
>>> y
False
>>> z = bool(-10)
>>> z
True
```

布尔变量也可以进行加、减、乘、除运算。需要注意的是,True 的值视为 1,False 的值视为 0。下面通过示例进行说明:

```
>>> x = True
>>> y = False
>>> x + y
1
>>> x - y
1
>>> x * y
0
>>> x / y
Traceback (most recent call last):
    File "<stdin>", line 1, in <module>
```

```
ZeroDivisionError: division by zero
>>> x = True
>>> y = True
>>> x + y
2
>>> x - y
0
>>> x * y
1
>>> x / y
1.0
>>> x = False
>>> y = False
>>> x + y
0
>>> x - y
0
>>> x * y
0
>>> x / y
Traceback(most recent call last):
    File "<stdin>", line 1, in <module>
ZeroDivisionError: division by zero
```

在大多数情况下,布尔变量的基本运算与处理 1 或 0 参与的运算结果一致。由于出现除数为零的情况会得到一个"ZeroDivisionError"的异常,所以要小心除数为零的情况。还可以使用以下语法创建字节、字节数组和内存视图对象[①]。

```
>>> x = b"Hello World"
>>> x
b'Hello World'
>>> y = bytearray(6)
>>> y
bytearray(b'\x00\x00\x00\x00\x00\x00')
>>> z = memoryview(bytes(5))
>>> z
<memory at 0x00000134C2A1D900>
```

可以用字符串拼接的方式将字节串拼接在一起。示例如下:

```
>>> x = b"Hello World"
>>> x
b'Hello World'
>>> y = b" My name is Rob"
>>> y
b' My name is Rob'
```

---

① 不同运行环境下,memoryview 得到的地址不同,所以示例运行结果可能不一致。

```
>>> x + y
b'Hello World My name is Rob'
```

本章小结

本章介绍了 Python 中的一些数据类型,以及如何使用数学方法对其进行操作。读者从中可以领悟 Python 对象的操作机制以及参数传递方法。

CHAPTER 5

第5章 运算符

前面章节介绍了数据类型以及一些基本的运算符,本章将继续介绍一些比较重要的运算符,理解起来也较为简单。与第4章的示例相似,可以定义如下变量:

```
>>> x = 2 + 1
>>> x
3
```

上述示例中,将表达式"2+1"的运算结果3赋值给变量x。现在,如果想要验证x的值是否为3,可以使用"=="进行相等性判断。

```
>>> x = 2 + 1
>>> x == 3
True
```

相等性可以用"=="运算符进行判断,那么如何判别不相等呢?Python使用"!="运算符判断不相等。使用前面的示例,得到以下结果:

```
>>> x = 2 + 1
>>> x != 3
False
```

在此得到的是一种比较结果,类型为布尔型(True或False)。

也可以测试对象是否大于或小于某一个值。

```
>>> x = 2 + 1
>>> x > 4
False
>>> x < 4
True
>>> x >= 4
False
>>> x >= 3
```

```
True
>>> x <= 3
True
```

在此,还存在以下比较运算符,将运算符左边的值和右边的值进行比较并测试,如表 5.1 所示。

表 5.1 比较运算符

运 算 符	说 明
>	大于
<	小于
>=	大于或等于
<=	小于或等于

is 语句也可以用于测试两个变量的相等性。is 语句与前面讲到的"=="运算符不尽相同,二者本质的区别在于关注的重点不同。如果所比较的两个变量指向同一对象,则 is 语句返回 True;如果两个对象的值相等,则"=="运算符返回 True。下述示例将展示 is 语句与"=="运算符的区别。

```
>>> a = 1
>>> b = 1
>>> a is b
True
>>> a == b
True
>>> a = []
>>> b = []
>>> a is b
False
>>> a == b
True
```

在前两条语句中①,分别将 1 赋值给变量 a 和变量 b,此时"a is b"返回 True。在后面的语句中,将空列表分别赋值给变量 a 和变量 b(列表的概念后面讲述),此时"a is b"返回的结果为 False,而"a == b"返回的结果为 True。产生这一结果的原因是 a 和 b 根本就不是同一个列表,因此会有:

```
>>> a is b
False
```

然而,a 和 b 都是列表,使"=="运算符比较等到的结果为 True,这是因为 a 和 b 都

---

① 译者注:本例译者进行了改动,作者给出的例子是"a=1, a is 1"将 a 与 1 直接用 is 判断会出错,原因是 is 不能用于直接与字面值进行比较。

是空列表,其值相等。如果将 a 赋值为空列表,然后令 b=a,使用 is 判断会出现如下结果:

```
>>> a = []
>>> b = a
>>> a is b
True
```

此时结果为 True,产生这一结果的原因是变量 b 与变量 a 相同,这两个变量具有相同的 id 值。与"=="和"!="一样都是测试两个变量的相等性,而 is 与 is not 是测试两个变量的身份是否相同。作为示例,可以将变量与值为 None 的 Python 空值(Null)进行比较,示例代码如下:

```
>>> a = 21
>>> a is not None
True
```

可以像前面一样,通过给变量赋值来覆盖变量原来的值。

```
>>> x = 1
>>> x
1
>>> x = 10
>>> x
10
```

现在,如果有 3 个变量要赋值,可以按如下方式进行:

```
>>> x = 1
>>> y = 2
>>> z = 3
>>> x
1
>>> y
2
>>> z
3
```

上述代码运行没有问题,但是占用了很多行,可以按如下方式进行简化赋值:

```
>>> x, y, z = 1, 2, 3
>>> x
1
>>> y
2
>>> z
3
```

在同一行为多个变量赋值的方式会使变量赋值更加容易,代码量缩短,程序可读性更高。显然,前面示例中为变量的命名习惯不是很好,如果能给变量取一个有意义的名称会让代码更有意义,可以增加代码的易读性。

前面已经讲解了如何为变量赋值,下面介绍如何修改变量的值和如何修改值的大小。假设有一个变量 profit,现在要将其加上 100.0,可以按照如下代码进行实现:

```
>>> profit = 1000.0
>>> profit
1000.0
>>> profit = profit + 100.0
>>> profit
1100.0
```

上述代码中,定义了变量 profit,并为其赋初始值,接着在其值上加了 100.0。这样编写程序没有什么问题,但可以写成更加 Python 化的形式:

```
>>> profit = 1000.0
>>> profit
1000.0
>>> profit += 100.0
>>> profit
1100.0
```

类似地,如果想要将 profit 变量的值增加 20%,可以用下述代码实现:

```
>>> profit = 1000.0
>>> profit
1000.0
>>> profit *= 1.2
>>> profit
1200.0
```

同样的方法也适用于除法和减法,示例代码如下:

```
>>> profit = 1000.0
>>> profit
1000.0
>>> profit -= 1.2
>>> profit
998.8
>>> profit /= 2
>>> profit
499.4
```

Python 中还可以使用一些非典型运算符。例如,取模运算符"%",它将返回整数除法的余数。示例如下:

```
>>> x = 2
>>> y = 10
>>> x / y
0.2
>>> x % y
2
```

下面是求幂运算,示例代码如下:

```
>>> x = 2
>>> y = 10
>>> y ** x
100
```

Python 还提供了地板除[①](floor division)运算符"//",该运算符在 Python2 时就开始使用。示例代码如下:

```
>>> x = 2
>>> y = 10
>>> x / y
0.2
>>> x // y
0
```

如前所述,可以使用"="运算符方法将运算结果赋值给变量。因此,可以按如下方式执行取模运算、求幂运算和地板除运算。

```
>>> x = 2
>>> y = 10
>>> x %= y
>>> x
2
>>> x = 2
>>> y = 10
>>> y **= x
>>> y
100
>>> x = 2
>>> y = 10
>>> x //= y
>>> x
0
```

还可以将这些类型的运算符复合在一起使用,如下所示:

---

① 地板除也称为向下取整,取的是除法结果的整数部分。

```
>>> x = 2
>>> y = 10
>>> x < y
True
>>> z = 3
>>> a = 3
>>> b = 3
>>> a == b
True
>>> x < y and a == b
True
>>> x > y
False
>>> x > y and a == b
False
>>> a != b
False
>>> x > y and a != b
False
```

因此,我们可以将逻辑语句组合在一起形成更复杂的语句。当将这些内容应用于本书后面的一些功能时才会真正显示出其价值。

本章小结

本章展示了如何对不同 Python 数据类型执行各种操作的方法,这将构成本书后续应用的逻辑基础。

CHAPTER 6

第 6 章 日 期

第 4 章绍了 Python 中的一些主要数据类型,但有一件事非常重要,那就是日期。对于处理过日期的人来说,日期类型有很多种格式,处理起来十分烦琐,也很难操作。但是 Python 中包含了日期数据类型的处理。如果想要创建一个 datetime 对象,可以按照如下方式进行操作(datetime 包的导入在第 3 章中已经介绍过):

```
>>> from datetime import datetime as dt
>>> d = dt(2017,5,17,12,10,11)
>>> d
datetime.datetime(2017, 5, 17, 12, 10, 11)
>>> str(d)
'2017-05-17 12:10:11'
```

上述代码中,创建了一个 datetime 对象,并将年、月、日、时、分、秒等参数传递给 dt() 函数,从而得到了 datetime 对象 d。如果想以更直观的方式进行查看,可以借助 str() 函数得到日期的字符串表示。如果想要计算日期,则可以按如下方式进行日期运算:

```
>>> d1 = dt(2017, 5, 17, 12, 10, 11)
>>> d1
datetime.datetime(2017, 5, 17, 12, 10, 11)
>>> d2 = dt(2016, 4, 7, 1, 1, 1)
>>> d2
datetime.datetime(2016, 4, 7, 1, 1, 1)
>>> d1 - d2
datetime.timedelta(days = 405, seconds = 40150)
>>> str(d1 - d2)
'405 days, 11:09:10'
```

上述代码中,首先创建了两个日期,然后将一个日期减去另一个日期,这样就得到了一个 timedelta 对象,接着将其转换为一个字符串,最后得到了日期减法之后的天数和时间。为了理解 timedelta,可以将其导入,与导入 datetime 一样。

```
>>> from datetime import timedelta as td
```

timedelta[①]与 datetime 略有不同,timedelta 是 datetime 模块中的一个方法,是一个时间间隔对象,表示时间长度。timedelta 中的时间间隔可以传入日、时、分和秒参数,而不能传入年、月参数。将年和月作为参数传入 timedelta 的意义不大,因为年和月并不是 timedelta 支持的统一时间单位。可以创建一个 1 天 2 小时 10 分钟的 timedelta 对象,如下所示:

```
>>> td(days = 1, hours = 2, minutes = 10)
datetime.timedelta(days = 1, seconds = 7800)
>>> change = td(days = 1, hours = 2, minutes = 10)
>>> d1 = dt(2017, 5, 17, 12, 10, 11)
>>> d1 - change
datetime.datetime(2017, 5, 16, 10, 0, 11)
>>> str(d1 - change)
'2017 - 05 - 16 10:00:11'
```

上述代码中,将 datetime 对象(d1)减去了 timedelta 对象(change),结果返回了减去 timedelta 时间长度的 datetime 数据。当它显示为字符串时,可以看到保留了 datetime 格式。因此,借助 timedelta 使做日期减法更加容易,如果想实现 datetime 加法也是如此,参考示例代码如下:

```
>>> str(d1 - change)
'2017 - 05 - 16 10:00:11'
>>> d1 + change
datetime.datetime(2017, 5, 18, 14, 20, 11)
>>> str(d1 + change)
'2017 - 05 - 18 14:20:11'
```

如果只想处理日期,则可以从 datetime 导入 date,示例代码如下:

```
>>> from datetime import date as d
>>> d.today()
datetime.date(2022, 10, 21)
>>> str(d.today())
'2022 - 10 - 21'
```

上述代码中,日期的导入方式与之前相同。假设想要当前日期,则可以使用 today() 方法显示当前日期。

利用上面所学内容,已经可以完成一些示例。例如,考虑给定一个时间日期,怎样计算这个日期距离现在的时间长度。假设给定的日期是 1969 年 7 月 20 日,即尼尔·阿姆

---

① timedelta 对象的用法可以参考 Python 文档(https://docs.python.org/zh-cn/3/library/datetime.html?highlight=timedelta#datetime.timedelta)。

斯特朗登上月球的日期，计算该日期距今有多远的示例代码如下：

```
>>> import datetime as dt
>>> now_date = dt.datetime.now()
>>> now_date
datetime.datetime(2022, 10, 21, 21, 51, 58, 469346)
>>> moon_date = dt.datetime(1969, 7, 20)
>>> moon_date
datetime.datetime(1969, 7, 20, 0, 0)
>>> date_since_moon = now_date - moon_date
>>> date_since_moon
datetime.timedelta(days = 19451, seconds = 78718, microseconds = 469346)
>>> date_since_moon.days
19451
>>> date_since_moon.seconds
78718
```

从上述代码可以看出，从一个日期中减去另一个日期会得到一个考虑了不同类型参数的 timedelta 时间差。

接下来，可以试着查看一个日期在未来有多远。例如，可以看一下 2030-01-01 这一日期相对于现在的时间差是多少（以当前程序执行时间为基准）：

```
>>> import datetime as dt
>>> now_date = dt.date.today()
>>> now_date
datetime.date(2022, 10, 21)
>>> future_date = dt.date(2030, 1, 1)
>>> future_date
datetime.date(2030, 1, 1)
>>> distance = future_date - now_date
>>> distance
datetime.timedelta(days = 2629)
>>> distance.days
2629
```

上述代码中，所做的是使用 today() 方法将日期设置为今天，然后减去设置为未来日期的日期，得到 timedelta 对象。如代码所示，可以算出 2022-10-21 和 2030-01-01 之间的天数为 2629 天。

本章小结

本章主要介绍了如何在 Python 中进行日期操作，展示了如何导入日期模块以及如何访问其中的方法和属性。

CHAPTER 7

第7章 列 表

本章将介绍列表。此处所说的列表不是在食品店随手记录的表格,而是 Python 的一个非常重要的内容。列表是 Python 中排在第一位的存储类型,其他类型是元组、字典和集合。列表最本质的属性是可以存储数据。例如,如果想要创建一个包含数字 1~10 的变量,可以将它们存储在列表中,如下所示:

```
>>> numbers = [1, 2, 3, 4, 5, 6, 7, 8, 9, 10]
>>> numbers
[1, 2, 3, 4, 5, 6, 7, 8, 9, 10]
```

上述代码中,列表的元素都是整数,同样方法可以创建一个由字符串构成的列表。

```
>>> numbers = ["1","2","3","4","5","6","7","8","9","10"]
>>> numbers
['1', '2', '3', '4', '5', '6', '7', '8', '9', '10']
```

列表的美妙之处在于列表可以包含多种数据类型,如下所示:

```
>>> stuff = ["longing","rusted","furnace","daybreak",17,"Benign",9]
>>> stuff
['longing', 'rusted', 'furnace', 'daybreak', 17, 'Benign', 9]
```

甚至可以将变量放在一个列表中。对于上个示例,可以用变量等价替换一些条目,如下所示:

```
>>> x = "longing"
>>> y = 17
>>> stuff = [x,"rusted","furnace","daybreak",y,"Benign",9]
>>> stuff
['longing', 'rusted', 'furnace', 'daybreak', 17, 'Benign', 9]
```

列表中也可以包含列表,示例如下:

```
>>> first_names = ["Steve", "Peter", "Tony", "Natasha"]
>>> last_names = ["Rodgers", "Parker", "Stark", "Romanoff"]
>>> names = [first_names, last_names]
>>> names
[['Steve', 'Peter', 'Tony', 'Natasha'], ['Rodgers', 'Parker', 'Stark', 'Romanoff']]
```

总体来说，列表的功能是非常强大的。但是如果在列表中放入了数据，就需要能够访问它。Python 中的列表是基于零索引（zero indexed）的，即列表索引从 0 开始，这意味着如果要访问列表的第 1 个元素，则需要执行以下操作：

```
>>> stuff = ["longing","rusted","furnace","daybreak",17,"Benign",9]
>>> stuff
['longing', 'rusted', 'furnace', 'daybreak', 17, 'Benign', 9]
>>> stuff[0]
'longing'
```

同样地，要访问列表的第 2 个元素和第 5 个元素，可以按照以下执行：

```
>>> stuff[1]
'rusted'
>>> stuff[4]
17
```

如果没有使用过零索引，则一开始可能会觉得有些混乱，但一旦掌握了零索引的诀窍，它就会让你跳出固定思维方式，释放编程方面的第二天性。在本章后续内容中将详细介绍如何访问列表。

访问列表的第一种方法是 pop()，该操作将删除列表的最后一项。

```
>>> stuff.pop()
9
>>> stuff
['longing', 'rusted', 'furnace', 'daybreak', 17, 'Benign']
```

利用 pop()，还可以指定要从列表中删除的元素的索引位置，如下所示：

```
>>> stuff
['longing', 'rusted', 'furnace', 'daybreak', 17, 'Benign']
>>> stuff.pop(4)
17
>>> stuff
['longing', 'rusted', 'furnace', 'daybreak', 'Benign']
```

需要注意的是，pop()方法返回的值是列表删除的元素值，不必将该值重新赋给列表，原先列表已经被修改。

使用 pop()方法删除末尾元素，可能让你觉得这很简单。这也很好地引导出访问列表的第二种方法 append()，该方法允许在列表的末尾添加一个元素。

```
>>> stuff.append(9)
>>> stuff
['longing', 'rusted', 'furnace', 'daybreak', 'Benign', 9]
```

还有一种方法可以从列表中删除列表元素,那就是 remove() 方法。假设我们想从列表中删除元素 9,可以使用以下方法:

```
>>> stuff
['longing', 'rusted', 'furnace', 'daybreak', 'Benign', 9]
>>> stuff.remove(9)
>>> stuff
['longing', 'rusted', 'furnace', 'daybreak', 'Benign']
```

remove() 方法可以删除列表中的任何元素,只需要指定要删除的元素名称。值得注意的是,remove() 方法不会删除列表的所有实例,只删除列表中该值的第一个实例。因此,可以再次使用 append() 方法将列表恢复到原来的样式。

```
>>> stuff.append(9)
>>> stuff
['longing', 'rusted', 'furnace', 'daybreak', 'Benign', 9]
```

接下来,将展示如何使用 count() 方法,借助该方法可以统计列表中等于某一数值的元素个数。为了说明这一点,下面将定义一个较为有趣的列表:

```
>>> count_list = [1,1,1,2,3,4,4,5,6,9,9,9]
>>> count_list.count(1)
3
>>> count_list.count(4)
2
```

现在,将展示如何使用 reverse() 方法,该方法可以将列表中的元素顺序进行反转。示例代码如下:

```
>>> count_list.reverse()
>>> count_list
[9, 9, 9, 6, 5, 4, 4, 3, 2, 1, 1, 1]
```

下面将展示最后一种访问列表方法,即 sort() 排序方法。对于一个整数列表来说,sort() 排序的工作方式如下所示:

```
>>> count_list = [9, 9, 9, 6, 5, 4, 4, 3, 2, 1, 1, 1]
>>> count_list.sort()
>>> count_list
[1, 1, 1, 2, 3, 4, 4, 5, 6, 9, 9, 9]
```

但是,如果列表内的元素使用字符串或混合数据类型,则可以借助前面定义的 stuff

列表查看执行结果。示例代码如下：

```
>>> stuff
['longing', 'rusted', 'furnace', 'daybreak', 'Benign', 9]
>>> stuff.sort()
Traceback(most recent call last):
    File "<stdin>", line 1, in <module>
TypeError: '<' not supported between instances of 'int' and 'str'
```

产生上述错误的原因是Python不知道如何对整数和字符串进行排序，因此我们只能对数值列表使用sort()排序方法。

此时，我们需要查看列表的属性，重新回到如何从列表中选择元素的问题。在前面引入了索引的概念，并根据索引选择列表中的元素。下面，将进一步展示当索引值为负数时的工作原理。简单地说，负索引是从列表的末尾开始，从后向前计数，所以列表的右边最后一个元素的索引是-1，列表左边第一个元素的索引是列表长度的负值。可以借助len()函数获得列表的长度，示例代码如下：

```
>>> stuff = ["longing", "rusted", "furnace", "daybreak", 17, "Benign", 9]
>>> len(stuff)
7
>>> stuff[-1]
9
>>> stuff[-2]
'Benign'
>>> stuff[-len(stuff)]
'longing'
```

上述代码中，首先将列表显示为之前定义的内容；然后将len()函数应用于列表并获得列表的长度为7，也就意味着该列表有7个元素；接下来使用负索引选择最后一个元素和倒数第二个元素；最后给出了如何使用负索引获取第1个元素的方法。

现在，从列表中选择一个元素已经不在话下了。如果我们想选择列表的子集，那该怎么办呢？Python支持这种操作。假设想选择第2个和第3个元素，并将结果显示在新的列表中，就可以使用前面讲述的方法来实现，示例代码如下：

```
>>> x = stuff[1]
>>> y = stuff[2]
>>> new_stuff = [x, y]
>>> new_stuff
['rusted', 'furnace']
```

这样实现虽然可行，但代码并不简洁，其实用一行代码就可以实现，如下所示：

```
>>> stuff[1:3]
['rusted', 'furnace']
```

上述代码中的索引区间是从列表中索引1(这是列表的第2个元素,因为索引从0开始)中开始提取元素,直到索引为3的元素截止,但不包括索引为3的元素。按照类似的方式,可以选择列表中除第1个元素以外的所有元素,示例代码如下:

```
>>> stuff[1:]
['rusted', 'furnace', 'daybreak', 17, 'Benign', 9]
>>> stuff
['longing', 'rusted', 'furnace', 'daybreak', 17, 'Benign', 9]
```

请注意,上述代码利用索引区间获取列表中所需内容之后显示了列表,列表依然是完整的。这是因为列表的切片操作(splicing)不会改变列表的内容,它只是返回列表的切片部分。因此,如果要利用切片操作将结果赋值给一个变量,可以参考如下示例代码:

```
>>> new_stuff = stuff[1:]
>>> new_stuff
['rusted', 'furnace', 'daybreak', 17, 'Benign', 9]
```

切片操作也可以使用负索引。如果使用-1而不是1,则得到以下结果:

```
>>> new_stuff = stuff[-1:]
>>> new_stuff
[9]
```

上述代码是想要从索引为-1的元素开始,一直到列表的末尾,把所选中的元素都保存下来。如果将索引放在冒号的后面,就可以形成新的切片。因此,将-1放在冒号的后面,可以将上述代码修改为:

```
>>> new_stuff = stuff[:-1]
>>> new_stuff
['longing', 'rusted', 'furnace', 'daybreak', 17, 'Benign']
```

因此,上述代码与之前示例相反,获取从列表起始元素开始一直到索引为-1的元素结束,但不包括索引值为-1的元素。

假设现在要在列表中每两个元素选取一个,则可以按照以下程序进行操作:

```
>>> stuff
['longing', 'rusted', 'furnace', 'daybreak', 17, 'Benign', 9]
>>> stuff[1:8:2]
['rusted', 'daybreak', 'Benign']
```

上述代码中,首先选择索引1和索引7之间的所有元素,然后从第一个位置开始每两个元素选取一个。这一功能非常强大,可以让我们对列表进行许多控制操作。

接下来,看一下如何将列表拼接在一起。前面已经介绍了字符串的拼接,列表的拼接操作与其非常相似,示例如下:

```
>>> stuff
['longing', 'rusted', 'furnace', 'daybreak', 17, 'Benign', 9]
>>> count_list
[1, 1, 1, 2, 3, 4, 4, 5, 6, 9, 9, 9]
>>> stuff + count_list
['longing', 'rusted', 'furnace', 'daybreak', 17, 'Benign', 9, 1, 1, 1, 2, 3, 4, 4, 5, 6, 9, 9, 9]
```

与字符串拼接类似,可以使用"+"符号将两个或多个列表拼接在一起。

如果想检查一个元素是否在列表中,可以借助 Python 中的 in 语句轻松做到这一点。如果元素在列表中,则返回布尔值 True,否则返回布尔值 False。

```
>>> stuff
['longing', 'rusted', 'furnace', 'daybreak', 17, 'Benign', 9]
>>> 9 in stuff
True
```

同样地,可以使用 not in 语句检查一个元素是否不在列表中,使用 not in 重复上述代码得到结果如下:

```
>>> stuff
['longing', 'rusted', 'furnace', 'daybreak', 17, 'Benign', 9]
>>> 9 not in stuff
False
```

现在,考虑列表的复制方法。为了展示这一功能,下面将创建一个列表,然后将其赋值给一个新列表。

```
>>> stuff = ["longing", "rusted", "furnace", "daybreak", 17, "Benign", 9]
>>> stuff
['longing', 'rusted', 'furnace', 'daybreak', 17, 'Benign', 9]
>>> new_stuff = stuff
>>> new_stuff
['longing', 'rusted', 'furnace', 'daybreak', 17, 'Benign', 9]
>>> stuff.append(21)
>>> stuff
['longing', 'rusted', 'furnace', 'daybreak', 17, 'Benign', 9, 21]
>>> new_stuff
['longing', 'rusted', 'furnace', 'daybreak', 17, 'Benign', 9, 21]
```

正如上述代码所示,以赋值方式创建第二个列表 new_stuff 之后,对 stuff 列表的任何操作都会反映在 new_stuff 列表中。如果不想这样做,只是想要复制列表(二者使用时具有独立性),可以使用 copy()方法。示例代码如下:

```
>>> stuff = ["longing", "rusted", "furnace", "daybreak", 17, "Benign", 9]
>>> stuff
['longing', 'rusted', 'furnace', 'daybreak', 17, 'Benign', 9]
```

```
>>> new_stuff = stuff.copy()
>>> new_stuff
['longing', 'rusted', 'furnace', 'daybreak', 17, 'Benign', 9]
>>> stuff.append(21)
>>> stuff
['longing', 'rusted', 'furnace', 'daybreak', 17, 'Benign', 9, 21]
>>> new_stuff
['longing', 'rusted', 'furnace', 'daybreak', 17, 'Benign', 9]
```

在本章中考虑的下一个方法是 clear() 方法,该方法将清除列表中的所有内容。

```
>>> stuff = ["longing", "rusted", "furnace", "daybreak", 17, "Benign", 9]
>>> stuff.clear()
>>> stuff
[]
```

下面将考虑的最后一种方法并不是严格的列表方法。这里引入 range 对象。在 Python2 中,可以使用以下语法创建一个 range 对象[①]。

```
>>> x = range(7)
>>> x
[0, 1, 2, 3, 4, 5, 6]
>>> type(x)
<type 'list'>
```

上述代码中,range() 函数创建了一个从 0 开始的长度为 7 的列表。可以通过给 range() 函数设置起点和终点来修改它,如下所示:

```
>>> x = range(1,7)
>>> x
[1, 2, 3, 4, 5, 6]
```

由上可见,列表从 1 开始,到 6 结束,这与之前使用冒号对列表进行切片操作类似。我们可以进一步增加第 3 个参数,如下所示:

```
>>> x = range(1,7,2)
>>> x
[1, 3, 5]
```

上述示例得到了一个从 1 开始到 6 结束并且间隔一个元素取值的列表。range() 函数有 3 个参数:start、stop 和 step,其中 stop 是唯一必选参数。range() 函数对于创建动态列表非常有用,但是在 Python 3 中进行了调整,range() 函数不再创建列表对象,而是创建了一个 range 对象。下面,利用 Python3 再演示一下前面的示例。

---

① Python3 中,range 是单独的类型,该代码在 Python3 中运行后,type 函数的结果为"<class 'range'>"。

```
>>> x = range(7)
>>> x
range(0, 7)
>>> type(x)
<class 'range'>
```

像以前一样,给 range() 函数传入 start 值和 stop 值。

```
>>> x = range(1,7)
>>> x
range(1, 7)
```

同样地,给 range() 函数增加 step 参数。

```
>>> x = range(1,7,2)
>>> x
range(1, 7, 2)
```

我们可以访问 range 对象的元素,就像通过索引值访问列表中的元素一样,示例代码如下:

```
>>> x = range(7)
>>> x[0]
0
>>> x[-1]
6
>>> x[3]
3
```

range 对象也支持切片,就像列表进行切片操作一样,并且可以访问其中的元素。

```
>>> x = range(7)
>>> x[1:]
range(1, 7)
>>> x[:-1]
range(0, 6)
>>> x[:-1][-1]
5
>>> x[3]
3
>>> x[:4:2]
range(0, 4, 2)
```

与列表不同,range 函数没有与之关联的各种方法,但可以获得 range 对象的 start、stop 和 step 以及长度和索引。

```
>>> x = range(7)
>>> x = x[1:]
```

```
>>> x
range(1, 7)
>>> x.start
1
>>> x.stop
7
>>> x.step
1
>>> x.index(1)
0
>>> x.count(1)
1
>>> x = range(7)
>>> x = x[1:5:2]
>>> x
range(1, 5, 2)
>>> x.start
1
>>> x.stop
5
>>> x.step
2
>>> x.index(3)
1
>>> x.count(3)
1
```

range()函数对于创建包含整数的动态对象非常有用,本书后面的一些示例中经常会用到。

本章小结

本章主要讲解了列表的概念,为后续如何创建、访问和操作列表提供了很好的参考,在后续章节中都会用到列表。

CHAPTER 8

第 8 章　元　　组

有了列表的知识,学习元组将会十分容易。元组本质上是列表,可以使用与访问列表相同的方式访问元组,列表的许多功能都可以应用到元组。元组与列表最大的区别是元组不能被修改。既然元组不能被修改,那它还有什么用呢？实际上,元组比你想象的更有用,借助元组可以防止意外更改代码中可能依赖的东西。

下述列表是在第 7 章开头定义的 numbers 列表：

```
>>> numbers = [1, 2, 3, 4, 5, 6, 7, 8, 9, 10]
>>> numbers
[1, 2, 3, 4, 5, 6, 7, 8, 9, 10]
```

将上述 numbers 修改为元组,示例代码如下：

```
>>> numbers = (1, 2, 3, 4, 5, 6, 7, 8, 9, 10)
>>> numbers
(1, 2, 3, 4, 5, 6, 7, 8, 9, 10)
```

可以看到,上述代码与前面的列表看起来十分相似。现在,试着访问元组的第 1 个元素(采用与列表相同的方式),示例代码如下：

```
>>> numbers[0]
1
```

同样,获取元组的最后一个元素：

```
>>> numbers[-1]
10
```

此外,还可以按照列表方法对元组进行切片操作。因此,如果想要获取元组中除去最后两个元素以外的其他元素,可以按照列表切片相同的方式获取,示例如下：

```
>>> numbers[:-2]
(1, 2, 3, 4, 5, 6, 7, 8)
```

那么，使用元组的意义是什么呢？为了解释该问题，让我们试着修改元组的第二个元素，就像操作列表一样，示例代码如下：

```
>>> numbers[1] = 1
Traceback(most recent call last):
    File "<stdin>", line 1, in <module>
TypeError: 'tuple' object does not support item assignment
```

此时，代码报错"元组对象不支持项分配"，即不能对一个已经定义的元组赋值。那么，我们能对元组做什么呢？在本例中，对于 numbers 元组，可以使用两个方法，即 count() 和 index()。

count() 方法用于返回元组中某个值出现的次数。为了更好地说明，下面将定义一个新的元组 new_numbers，如下所示：

```
>>> new_numbers = (1, 2, 2, 2, 5, 5, 7, 9, 9, 10)
>>> new_numbers.count(2)
3
```

上述代码结果表明，new_numbers 元组中包含数值为 2 的实例有 3 个。

index() 方法一般用于检索并返回元组中某个值第一次出现的索引。示例如下：

```
>>> new_numbers = (1, 2, 2, 2, 5, 5, 7, 9, 9, 10)
>>> new_numbers.index(2)
1
```

上述示例展示了如何检索数值 2 在 new_numbers 元组中第一次出现的索引值，结果返回 1（这是正确的）。可以看到，在 new_numbers 元组中，索引值 2 和索引值 3 处也有元素 2。因此需要注意：index() 方法仅返回找到的第一个元素的索引值。

本章小结

本章简要介绍了元组的概念。实际上，在 Python 中拥有一个无法修改的对象是非常有用的，而元组正好具有这一特性，它在 Python 的许多包中都有使用。

CHAPTER 9

第 9 章　字　　典

当在 Python 中提起字典（dictionary）时，并不是在说牛津字典的变体。字典用于存放具有映射关系的数据，数据的存储形式为"键值对（key value pairs）"。"键值对"是什么意思呢？例如，在字典中，可以定义键（key）为 first name，对应的值（value）为 Rob；也可以定义键（key）为 wins，对应的值（value）为 21。字典的关键在于，如果要访问一个值，就需要有一个与之关联的键。下面以个人信息为例进行说明。

假设有个人信息字段：first name、surname、gender、favourite food，用字典的形式描述该字段，示例代码如下：

```
>>> dict_detail = {}
>>> dict_detail["first name"] = "Rob"
>>> dict_detail["surname"] = "Mastrodomenico"
>>> dict_detail["gender"] = "Male"
>>> dict_detail["favourite food"] = "Pizza"
>>> dict_detail
{'first name': 'Rob', 'surname': 'Mastrodomenico', 'gender': 'Male', 'favourite food': 'Pizza'}
>>>
```

需要注意的是，字典的名称可以是任意的。上述代码用"dict_detail"表示字典，此外，还可以很容易地将字典名称写成如下形式：

```
>>> ham_sandwich = {}
>>> ham_sandwich["first name"] = "Rob"
>>> ham_sandwich["surname"] = "Mastrodomenico"
>>> ham_sandwich["gender"] = "Male"
>>> ham_sandwich["favourite food"] = "Pizza"
>>> ham_sandwich
{'first name': 'Rob', 'surname': 'Mastrodomenico', 'gender': 'Male', 'favourite food': 'Pizza'}
>>>
```

对于上述字典名称的命名，你可能会认为第二个名称（即"ham_sandwich"）更有趣，这一点我也同意。实际上，为了增强代码的可读性，在命名字典或变量的名称时应该反映出想要做的事情。dict()是 Python 的内置函数，可以借助该函数完成想要的功能。但是，如果用自定义字典 dict()覆盖了内置函数 dict()，则就不能在当前环境中使用它。接下来请看下面的示例：

```
>>> person_details = dict(first_name = "Rob", surname = "Mastrodomenico",
... gender = "Male", favourite_food = "Pizza")
>>> person_details
{'first_name': 'Rob', 'surname': 'Mastrodomenico', 'gender': 'Male', 'favourite_food': 'Pizza'}
```

可以观察到，上述代码中的字段 favorite food 被改名为 favorite_food。原因是 Python 会将两个单词之间的空格解释为两个单独的条目，并会抛出一个"异常"。因此，在本章的剩余部分中，会把 favorite food 都写成 favorite_food。

因此，重新创建 person_details 字典，得到以下结果：

```
>>> personal_details = {}
>>> personal_details["first name"] = "Rob"
>>> personal_details["surname"] = "Mastrodomenico"
>>> personal_details["gender"] = "Male"
>>> personal_details["favourite_food"] = "Pizza"
>>> personal_details
{'first name': 'Rob', 'surname': 'Mastrodomenico', 'gender': 'Male', 'favourite_food': 'Pizza'}
```

上述代码中的第 1 行表示创建字典，如下所示：

```
>>> personal_details = {}
```

后续代码行是对字典的键值对进行赋值。我们也可以使用另外一种方式对字典进行赋值，如下所示：

```
>>> personal_details = {"first name": "Rob", "surname": "Mastrodomenico", "gender": "Male", "favourite_food": "Pizza"}
>>> personal_details
{'first name': 'Rob', 'surname': 'Mastrodomenico', 'gender': 'Male', 'favourite_food': 'Pizza'}
```

类似地，我们可以利用 dict()函数实现与上述字典相同的结果，如下所示：

```
>>> personal_details = dict([("first name", "Rob"),("surname", "Mastrodomenico"),
("gender", "Male"),("favourite_food", "Pizza")])
>>> personal_details
{'first name': 'Rob', 'surname': 'Mastrodomenico', 'gender': 'Male', 'favourite_food': 'Pizza'}
```

前面讨论了"异常"的处理，下面简要介绍一下这部分内容。假设我们试图访问字典中不存在的键，如下所示：

```
>>> personal_details = dict([("first name", "Rob"),("surname", "Mastrodomenico"),
("gender", "Male"),("favourite_food", "Pizza")])
```

```
>>> personal_details
{'first name': 'Rob', 'surname': 'Mastrodomenico', 'gender': 'Male', 'favourite_food': 'Pizza'}
>>> personal_details["age"]
Traceback(most recent call last):
    File "<stdin>", line 1, in <module>
KeyError: 'age'
```

我们希望能够允许我们尝试去访问一个字典中不存在的键,而不是抛出一个异常(或错误)。从代码返回结果中可以非常明确地看到 Python 抛出了一个错误"KeyError",如果能对其进行处理,示例代码如下:

```
>>> personal_details = dict([("first name", "Rob"),("surname", "Mastrodomenico"),("gender", "Male"),("favourite_food", "Pizza")])
>>> personal_details
{'first name': 'Rob', 'surname': 'Mastrodomenico', 'gender': 'Male', 'favourite_food': 'Pizza'}
>>> try:
...     age = personal_details["age"]
... except KeyError:
...     age = None
...
>>> age
>>>
```

上述代码中,使用了 try-except 语句,将想要运行的代码放入 try 结构中,如果代码有错误不能执行,则它会处理错误。在该情况下,程序首先尝试访问 age 键,并将其对应的值赋值给变量 age,如果访问 age 键不成功,此时会抛出一个错误,当出现的错误恰好是"KeyError"时就将 None(空值)赋值给 age 变量。由此可见,该方法不但可以抛出错误,而且还能以更好的方式处理无法找到该键的情况。虽然 try-except 方法非常有用,但是如果处理不好也会比较危险。如果不需要指定错误类型,则可以将代码重写为:

```
>>> personal_details = dict([("first name", "Rob"),("surname", "Mastrodomenico"),("gender", "Male"),("favourite_food", "Pizza")])
>>> personal_details
{'first name': 'Rob', 'surname': 'Mastrodomenico', 'gender': 'Male', 'favourite_food': 'Pizza'}
>>> try:
...     age = personal_details["age"]
... except:
...     age = None
...
>>> age
>>>
```

尽管 try-except 方法处理异常非常有用,但是如果有一种方法能获取字典中的键,且当该键不存在时也不会报错,那就更好了。幸运的是,Python 提供了一种 get()方法,它可以通过键来获取在字典中对应的值,即使键不存在也不会报错。

```
>>> personal_details = dict([("first name", "Rob"),("surname",
..."Mastrodomenico"),("gender", "Male"),("favourite_food",
..."Pizza")])
>>> personal_details
{'first name': 'Rob', 'surname': 'Mastrodomenico', 'gender': 'Male', 'favourite_food': 'Pizza'}
>>> personal_details.get('gender')
'Male'
>>> personal_details.get('age')
```

在第 7 章学习列表时，介绍了 pop() 方法，对于字典来说，也有类似的方法。示例代码如下：

```
>>> personal_details = dict([("first name","Rob"),("surname","Mastrodomenico"),
("gender","Male"),("favourite_food","Pizza")])
>>> personal_details
{'first name': 'Rob', 'surname': 'Mastrodomenico', 'gender': 'Male', 'favourite_food': 'Pizza'}
>>> personal_details.pop('gender')
'Male'
>>> personal_details
{'first name': 'Rob', 'surname': 'Mastrodomenico', 'favourite_food': 'Pizza'}
>>> personal_details.popitem()
('favourite_food', 'Pizza')
>>> personal_details
{'first name': 'Rob', 'surname': 'Mastrodomenico'}
```

上述示例中，使用了两种方法删除字典中的键值对。第一种方法是 pop() 方法，利用该方法从字典中删除了指定的键及其对应的值。第二种方法使用的是 popitem() 方法，该方法只删除字典中的最后一个键值对。

还有一种方法可以删除字典中的元素，那便是 del()。将想要删除的键和其所在的字典名称传递给 del() 函数，便可以从字典中删除键值对。

```
>>> personal_details = dict([("first name","Rob"),("surname","Mastrodomenico"),
("gender","Male"),("favourite_food","Pizza")])
>>> personal_details
{'first name': 'Rob', 'surname': 'Mastrodomenico', 'gender': 'Male', 'favourite_food': 'Pizza'}
>>> del personal_details['gender']
>>> personal_details
{'first name': 'Rob', 'surname': 'Mastrodomenico', 'favourite_food': 'Pizza'}
```

前面我们介绍了可以将一个列表赋值给另一个列表，对一个列表进行修改则另一个列表也会出现同样变化，对于字典也是如此。

```
>>> personal_details = dict([("first name","Rob"),("surname","Mastrodomenico"),
("gender","Male"),("favourite_food","Pizza")])
>>> personal_details
{'first name': 'Rob', 'surname': 'Mastrodomenico', 'gender': 'Male', 'favourite_food': 'Pizza'}
>>> his_details = personal_details
```

```
>>> his_details
{'first name': 'Rob', 'surname': 'Mastrodomenico', 'gender': 'Male', 'favourite_food': 'Pizza'}
>>> personal_details['age'] = 24
>>> personal_details
{'first name': 'Rob', 'surname': 'Mastrodomenico', 'gender': 'Male', 'favourite_food': 'Pizza', 'age': 24}
>>> his_details
{'first name': 'Rob', 'surname': 'Mastrodomenico', 'gender': 'Male', 'favourite_food': 'Pizza', 'age': 24}
```

如果想要复制一个字典，并可以独立地对其进行修改，可以使用copy()方法（与列表的copy()方法一样）。

```
>>> personal_details = dict([("first name", "Rob"),("surname", "Mastrodomenico"),("gender", "Male"),("favourite_food", "Pizza")])
>>> personal_details
{'first name': 'Rob', 'surname': 'Mastrodomenico', 'gender': 'Male', 'favourite_food': 'Pizza'}
>>> his_details = personal_details.copy()
>>> his_details
{'first name': 'Rob', 'surname': 'Mastrodomenico', 'gender': 'Male', 'favourite_food': 'Pizza'}
>>> personal_details['age'] = 24
>>> personal_details
{'first name': 'Rob', 'surname': 'Mastrodomenico', 'gender': 'Male', 'favourite_food': 'Pizza', 'age': 24}
>>> his_details
{'first name': 'Rob', 'surname': 'Mastrodomenico', 'gender': 'Male', 'favourite_food': 'Pizza'}
```

借助clear()方法，可以将字典中的所有元素都删除掉。

```
>>> personal_details = dict([("first name", "Rob"),("surname", "Mastrodomenico"),("gender", "Male"),("favourite_food", "Pizza")])
>>> personal_details
{'first name': 'Rob', 'surname': 'Mastrodomenico', 'gender': 'Male', 'favourite_food': 'Pizza'}
>>> personal_details.clear()
>>> personal_details
{}
```

在本章前面的内容中，我们讨论了创建字典的方法，但是如果我们想利用一个由键构成的元组来创建具有相同值的字典，可以使用fromkeys()方法。

```
>>> x = ('key1', 'key2', 'key3')
>>> y = 0
>>> res = dict.fromkeys(x, y)
>>> res
{'key1': 0, 'key2': 0, 'key3': 0}
```

到目前为止，我们已经能够使用键访问字典中的值。借助下面示例中给出的items()、keys()和values()方法，可以访问字典中的所有键和值。

```
>>> personal_details = dict([("first name", "Rob"),("surname", "Mastrodomenico"),
("gender", "Male"),("favourite_food", "Pizza")])
>>> personal_details
{'first name': 'Rob', 'surname': 'Mastrodomenico', 'gender': 'Male', 'favourite_food': 'Pizza'}
>>> personal_details.items()
dict_items([('first name', 'Rob'),('surname', 'Mastrodomenico'),('gender', 'Male'),
('favourite_food', 'Pizza')])
>>> personal_details.keys()
dict_keys(['first name', 'surname', 'gender', 'favourite_food'])
>>> personal_details.values()
dict_values(['Rob', 'Mastrodomenico', 'Male', 'Pizza'])
```

代码中返回的对象可以被迭代,稍后在引入循环时会介绍这一点。但是,如果想像访问列表一样访问字典,可以将其转换为列表,并借助索引进行相关访问。

```
>>> personal_details = dict([("first name", "Rob"),("surname", "Mastrodomenico"),
("gender", "Male"),("favourite_food", "Pizza")])
>>> personal_details
{'first name': 'Rob', 'surname': 'Mastrodomenico', 'gender': 'Male', 'favourite_food': 'Pizza'}
>>> personal_details.items()
dict_items([('first name', 'Rob'),('surname', 'Mastrodomenico'),('gender', 'Male'),
('favourite_food', 'Pizza')])
>>> list(personal_details.items())[0]
('first name', 'Rob')
>>> personal_details.keys()
dict_keys(['first name', 'surname', 'gender', 'favourite_food'])
>>> list(personal_details.keys())[-1]
'favourite_food'
>>> personal_details.values()
dict_values(['Rob', 'Mastrodomenico', 'Male', 'Pizza'])
>>> list(personal_details.values())[:-1]
['Rob', 'Mastrodomenico', 'Male']
```

本章小结

本章对字典的概念及其用法进行了介绍。与列表和元组相比,字典对象是一种非常灵活并且功能强大的容器,本书后面的很多内容都涉及字典的使用,足以说明字典概念在 Python 中的重要性。

CHAPTER 10

第 10 章 集 合

本章将介绍组合数据类型的最后一个对象,即集合(set)。集合中的数据不能重复,因此可以用于存储唯一的值。集合中的数据是无序的,并且不能被更改。下面开始创建一个集合:

```
>>> names = {'Tony','Peter','Natasha','Wanda'}
>>> names
{'Tony', 'Peter', 'Wanda', 'Natasha'}
```

上述代码中,我们使用花括号创建集合,与创建字典一样,但是没有键值对,内容类似于元组和列表。集合中的内容不仅限于字符串,可以在集合中添加一些整数,如下所示:

```
>>> names = {'Tony','Peter','Natasha','Wanda', 1, 2, 3}
>>> names
{1, 2, 3, 'Wanda', 'Natasha', 'Tony', 'Peter'}
```

可以看到,集合中元素的排序与我们放入的顺序不同。创建一个集合还可以借助前面介绍的 set() 内置函数。

```
>>> names = set(('Tony','Peter','Natasha','Wanda', 1, 2, 3))
>>> names
{1, 2, 3, 'Wanda', 'Natasha', 'Peter', 'Tony'}
>>> names = set(['Tony','Peter','Natasha','Wanda', 1, 2, 3])
>>> names
{1, 2, 3, 'Wanda', 'Natasha', 'Peter', 'Tony'}
```

借助字符串也可以创建集合对象,但需要知道使用花括号和 set() 内置函数的工作原理,如下所示:

```
>>> names = {'Wanda'}
>>> names
```

```
{'Wanda'}
>>> names = set('Wanda')
>>> names
{'W', 'd', 'n', 'a'}
```

上述程序中，当使用花括号传入字符串创建集合时，将保留完整的字符串，但当使用set()传入字符串时，字符串将被拆分为单个字符。需要注意的是，当字符串被分割时，并没有对字符进行排序。

接下来，我们尝试将列表添加到集合中，由于在集合中无法添加列表，因此引发了类型错误。

```
>>> my_set = {'Tony','Wanda', 1, 2, ['hello','world']}
Traceback (most recent call last):
    File "<stdin>", line 1, in <module>
TypeError: unhashable type: 'list'
```

对于字典和集合也是一样，也不能在集合中添加字典或者集合，如下所示：

```
>>> my_set = {'Tony','Wanda', 1, 2, {'key':'value'}}
Traceback (most recent call last):
    File "<stdin>", line 1, in <module>
TypeError: unhashable type: 'dict'
>>> my_set = {'Tony','Wanda', 1, 2, {1,2,3}}
Traceback (most recent call last):
    File "<stdin>", line 1, in <module>
TypeError: unhashable type: 'set'
```

然而，在集合中可以添加元组作为元素，如下所示：

```
>>> my_set = {'Tony','Wanda', 1, 2,(1,2,3)}
>>> my_set
{1, 2, 'Tony', 'Wanda',(1, 2, 3)}
```

在字典、列表和集合中可以添加元组的原因主要是元组的不可变性，因此在集合中得到支持。

可以使用下述语法查看某个值是否在集合中[①]：

```
>>> names = {'Tony','Peter','Natasha','Wanda', 1, 2, 3}
>>> 'Tony' in names
True
>>> 'Steve' in names
False
```

---

① 集合对象names的内容在前面代码中已经被修改，为了便于阅读，译者在下述代码中添加了一条语句重新给names赋值。

借助 add() 方法可以为集合添加元素。

```
>>> names.add('Steve')
>>> names
{1, 2, 3, 'Steve', 'Wanda', 'Natasha', 'Tony', 'Peter'}
>>> names.add('Tony')
>>> names
{1, 2, 3, 'Steve', 'Wanda', 'Natasha', 'Tony', 'Peter'}
```

请注意，当再次将 Tony 添加到集合中时，在集合中不会重复添加 Tony，但 Steve 会被添加，因为它原来不在集合中。在集合中不存在重复值这一特性非常有用，比如在可能有重复值的地方就可以用上集合的这一特性。例如，在一个列表中包含了大量的重复值，现在只需要其中的唯一值，我们就可以用集合的这一特性来解决，如下所示：

```
>>> days = ['Monday', 'Monday', 'Tuesday', 'Wednesday', 'Sunday', 'Sunday']
>>> days_set = set(days)
>>> days_set
{'Sunday', 'Wednesday', 'Monday', 'Tuesday'}
```

上面这个例子相当简单，因为可以看到列表中的所有内容，但可以想象如果是一个很大的数据集的情形，此时就会有类似的表示。

上述方法对于单个集合很有用，当有多个集合的时候，依然可以求取多个集合的唯一值。下面的示例将获取两个集合之间的唯一值。

```
>>> names = {'Tony','Peter','Natasha','Wanda'}
>>> names
{'Tony', 'Peter', 'Wanda', 'Natasha'}
>>> more_names = {'Steve', 'Peter', 'Carol', 'Wanda'}
>>> more_names
{'Carol', 'Peter', 'Wanda', 'Steve'}
>>> names | more_names
{'Wanda', 'Natasha', 'Peter', 'Steve', 'Carol', 'Tony'}
```

上面示例中，使用了"|"运算符，获取 names 集合或 more_names 集合中的值，所有的重复值只能包含一次，该功能同样可以使用集合的 union() 方法实现。

```
>>> names = {'Tony','Peter','Natasha','Wanda'}
>>> names
{'Tony', 'Peter', 'Wanda', 'Natasha'}
>>> more_names = {'Steve', 'Peter', 'Carol', 'Wanda'}
>>> more_names
{'Carol', 'Peter', 'Wanda', 'Steve'}
>>> names.union(more_names)
{'Wanda', 'Natasha', 'Peter', 'Steve', 'Carol', 'Tony'}
```

现在，可以看到上述两种方法的结果是一样的。如果不需要在 union() 方法中传递集合，则可以改用在union()方法中传递列表的方法来实现，如下所示：

```
>>> names = {'Tony','Peter','Natasha','Wanda'}
>>> names
{'Tony', 'Peter', 'Wanda', 'Natasha'}
>>> more_names = ['Steve', 'Peter', 'Carol', 'Wanda']
>>> more_names
['Steve', 'Peter', 'Carol', 'Wanda']
>>> names.union(more_names)
{'Wanda', 'Natasha', 'Peter', 'Steve', 'Carol', 'Tony'}
```

因此，当采用union()方法时，如果将集合转换为列表，也可以获得相同的结果。然而，使用"|"运算符时，如果将集合转换成列表，则程序会抛出一个错误。示例代码如下：

```
>>> names = {'Tony','Peter','Natasha','Wanda'}
>>> names
{'Tony', 'Peter', 'Wanda', 'Natasha'}
>>> more_names = ['Steve', 'Peter', 'Carol', 'Wanda']
>>> more_names
['Steve', 'Peter', 'Carol', 'Wanda']
>>> names | more_names
Traceback (most recent call last):
    File "<stdin>", line 1, in <module>
TypeError: unsupported operand type(s) for |: 'set' and 'list'
```

在前述union()示例的基础上，可以将更多项传递到union()方法中。例如，如果有两个集合，我们希望利用union()方法将这两个集合与现有集合合并，则可以执行以下操作：

```
>>> names = {'Tony','Peter','Natasha','Wanda'}
>>> names
{'Tony', 'Peter', 'Wanda', 'Natasha'}
>>> more_names = {'Steve', 'Peter', 'Carol', 'Wanda'}
>>> more_names
{'Carol', 'Peter', 'Wanda', 'Steve'}
>>> even_more_names = {'Tony', 'Jonny', 'Sue', 'Wade'}
>>> even_more_names
{'Wade', 'Tony', 'Sue', 'Jonny'}
>>> names | more_names | even_more_names
{'Wade', 'Wanda', 'Natasha', 'Peter', 'Steve', 'Sue', 'Jonny', 'Carol', 'Tony'}
>>> names.union(more_names, even_more_names)
{'Wade', 'Wanda', 'Natasha', 'Peter', 'Steve', 'Sue', 'Jonny', 'Carol', 'Tony'}
```

现在，如果传递的数据是列表而非集合，则依然可以使用union()方法进行数据合并，如下所示：

```
>>> names = {'Tony','Peter','Natasha','Wanda'}
>>> names
```

```
{'Tony', 'Peter', 'Wanda', 'Natasha'}
>>> more_names = ['Steve', 'Peter', 'Carol', 'Wanda']
>>> more_names
['Steve', 'Peter', 'Carol', 'Wanda']
>>> even_more_names = ['Tony', 'Jonny', 'Sue', 'Wade']
>>> even_more_names
['Tony', 'Jonny', 'Sue', 'Wade']
>>> names.union(more_names, even_more_names)
{'Wade', 'Wanda', 'Natasha', 'Peter', 'Steve', 'Sue', 'Jonny', 'Carol', 'Tony'}
```

到目前为止,这些示例已经研究了两个或多个集合之间的并集。如果我们想查看所有集合中的值,或者查看某值是否不在任何集合之中,Python 都可以进行处理。

前面已经介绍了如何利用"|"运算符和 union() 方法获取集合之间的唯一值。如果想要找出在所有集合中都有的值(交集),可以使用"&"运算符或 intersection() 方法。示例代码如下:

```
>>> names = {'Tony','Peter','Natasha','Wanda'}
>>> names
{'Tony', 'Peter', 'Wanda', 'Natasha'}
>>> more_names = {'Steve', 'Peter', 'Carol', 'Wanda'}
>>> more_names
{'Carol', 'Peter', 'Wanda', 'Steve'}
>>> names & more_names
{'Peter', 'Wanda'}
>>> names.intersection(more_names)
{'Peter', 'Wanda'}
```

正如在"|"运算符和 union() 方法中的用法一样,求取交集操作中也可以包含两个以上的集合。示例代码如下:

```
>>> names = {'Tony','Peter','Natasha','Wanda'}
>>> names
{'Tony', 'Peter', 'Wanda', 'Natasha'}
>>> more_names = {'Steve', 'Peter', 'Carol', 'Wanda'}
>>> more_names
{'Carol', 'Peter', 'Wanda', 'Steve'}
>>> even_more_names = {'Peter', 'Jonny', 'Sue', 'Wade'}
>>> even_more_names
{'Wade', 'Peter', 'Jonny', 'Sue'}
>>> names & more_names & even_more_names
{'Peter'}
>>> names.intersection(more_names, even_more_names)
{'Peter'}
```

与 union() 方法类似,我们可以将非集合数据添加到 intersection() 方法中。示例代码如下:

```
>>> names = {'Tony','Peter','Natasha','Wanda'}
>>> names
{'Tony', 'Peter', 'Wanda', 'Natasha'}
>>> more_names = ['Steve', 'Peter', 'Carol', 'Wanda']
>>> more_names
['Steve', 'Peter', 'Carol', 'Wanda']
>>> even_more_names = ['Peter', 'Jonny', 'Sue', 'Wade']
>>> even_more_names
['Peter', 'Jonny', 'Sue', 'Wade']
>>> names.intersection(more_names, even_more_names)
{'Peter'}
```

如果想查看两个或多个集合之间的差异,可以使用 difference() 方法或"-"运算符。同样,这里用到的规则与之前介绍的是一致的,只能使用"-"运算符处理集合数据,而 difference() 方法可以处理非集合数据。示例代码如下:

```
>>> names = {'Tony','Peter','Natasha','Wanda'}
>>> names
{'Tony', 'Peter', 'Wanda', 'Natasha'}
>>> more_names = {'Steve', 'Peter', 'Carol', 'Wanda'}
>>> more_names
{'Carol', 'Peter', 'Wanda', 'Steve'}
>>> names - more_names
{'Tony', 'Natasha'}
>>> names.difference(more_names)
{'Tony', 'Natasha'}
>>> even_more_names = {'Peter', 'Jonny', 'Sue', 'Wade'}
>>> even_more_names
{'Wade', 'Peter', 'Jonny', 'Sue'}
>>> names - more_names - even_more_names
{'Tony', 'Natasha'}
>>> names.difference(more_names, even_more_names)
{'Tony', 'Natasha'}
>>> more_names = ['Steve', 'Peter', 'Carol', 'Wanda']
>>> more_names
['Steve', 'Peter', 'Carol', 'Wanda']
>>> even_more_names = ['Peter', 'Jonny', 'Sue', 'Wade']
>>> even_more_names
['Peter', 'Jonny', 'Sue', 'Wade']
>>> names.difference(more_names, even_more_names)
{'Natasha', 'Tony'}
```

对多个集合比较差异的方式是从左到右,因此首先查看 names 和 more_names 之间的差异,随后查看该结果与 even_more_name 之间的差异。

另一种集合比较使用"^"运算符或 symmetric_difference() 方法,该运算返回两个集合中不重复的元素集合。演示示例如下:

```
>>> names = {'Tony','Peter','Natasha','Wanda'}
>>> names
{'Wanda', 'Tony', 'Peter', 'Natasha'}
>>> more_names = {'Steve', 'Peter', 'Carol', 'Wanda'}
>>> more_names
{'Steve', 'Carol', 'Wanda', 'Peter'}
>>> names ^ more_names
{'Tony', 'Natasha', 'Steve', 'Carol'}
>>> names.symmetric_difference(more_names)
{'Tony', 'Natasha', 'Steve', 'Carol'}
>>> even_more_names = {'Peter', 'Jonny', 'Sue', 'Wade'}
>>> even_more_names
{'Jonny', 'Sue', 'Wade', 'Peter'}
>>> names ^ more_names ^ even_more_names
{'Jonny', 'Wade', 'Tony', 'Natasha', 'Sue', 'Peter', 'Steve', 'Carol'}
>>> names.symmetric_difference(more_names, even_more_names)
Traceback(most recent call last):
    File "<stdin>", line 1, in <module>
TypeError: set.symmetric_difference() takes exactly one argument(2 given)
```

与前面的方法不同,symmetric_difference()方法不允比较多个集合,但它仍然允许我们传入非集合参数,如下所示:

```
>>> names = {'Tony','Peter','Natasha','Wanda'}
>>> names
{'Wanda', 'Natasha', 'Peter', 'Tony'}
>>> more_names = ['Steve', 'Peter', 'Carol', 'Wanda']
>>> more_names
['Steve', 'Peter', 'Carol', 'Wanda']
>>> names.symmetric_difference(more_names)
{'Natasha', 'Tony', 'Steve', 'Carol'}
```

我们还可以使用isdisjoint()方法判断两个集合是否包含相同的元素,如下所示:

```
>>> names = {'Tony','Peter','Natasha','Wanda'}
>>> names
{'Wanda', 'Natasha', 'Peter', 'Tony'}
>>> more_names = {'Steve', 'Bruce', 'Carol', 'Wanda'}
>>> more_names
{'Steve', 'Carol', 'Wanda', 'Bruce'}
>>> names.isdisjoint(more_names)
False
>>> more_names = {'Steve', 'Bruce', 'Carol', 'Sue'}
>>> more_names
{'Steve', 'Carol', 'Sue', 'Bruce'}
>>> names.isdisjoint(more_names)
True
```

如果读者对集合理论感兴趣,还有一对其他的集合方法可供使用:

- issubset()；
- issuperset()。

我们将这两个方法留给读者去研究它们的用法。

现在，除了这些集合比较的方法之外，还有许多可以应用于集合的方法，其中一些已经在前面的章节中介绍过，下面对其用法进行简要说明。

```
>>> names = {'Steve', 'Wanda', 'Peter', 'Tony', 'Natasha'}
>>> names
{'Peter', 'Steve', 'Wanda', 'Natasha', 'Tony'}
>>> names.pop()
'Peter'
```

如前所述，pop()方法从集合中删除了一个元素。

我们还可以使用remove()方法显式地从集合中移除元素：

```
>>> names = {'Steve', 'Wanda', 'Peter', 'Tony', 'Natasha'}
>>> names
{'Peter', 'Steve', 'Wanda', 'Natasha', 'Tony'}
>>> names.remove('Tony')
>>> names
{'Peter', 'Steve', 'Wanda', 'Natasha'}
```

如果试图使用remove()方法从集合中删除不存在的元素，会返回KeyError错误。另一种从集合中删除某个元素的方法是discard()方法，该方法允许删除不在集合中的元素，如下所示：

```
>>> names = {'Steve', 'Wanda', 'Peter', 'Tony', 'Natasha'}
>>> names
{'Peter', 'Steve', 'Wanda', 'Natasha', 'Tony'}
>>> names.remove('Sue')
Traceback(most recent call last):
    File "<stdin>", line 1, in <module>
KeyError: 'Sue'
>>> names.discard('Sue')
>>> names
{'Peter', 'Steve', 'Wanda', 'Natasha', 'Tony'}
>>> names.discard('Peter')
>>> names
{'Steve', 'Wanda', 'Natasha', 'Tony'}
```

借助clear()方法可以清除集合中的所有元素，得到一个空的集合。

```
>>> names = {'Steve', 'Wanda', 'Peter', 'Tony', 'Natasha'}
>>> names.clear()
>>> names
set()
```

如果从集合中移除一个元素就像将元素添加到集合中一样,我们用到的第一个方法就是 add() 方法。示例代码如下:

```
>>> names = {'Steve', 'Wanda', 'Peter', 'Tony', 'Natasha'}
>>> names
{'Peter', 'Tony', 'Wanda', 'Steve', 'Natasha'}
>>> names.add('Bruce')
>>> names
{'Peter', 'Bruce', 'Tony', 'Wanda', 'Steve', 'Natasha'}
>>> names.add('Peter')
>>> names
{'Peter', 'Bruce', 'Tony', 'Wanda', 'Steve', 'Natasha'}
```

可以看出,使用 add() 方法一次可以添加一个值,但可以使用一些性质上与集合比较类似的允许修改集合的方法,我们将展示的第一个方法是 update() 方法。示例代码如下:

```
>>> names = {'Tony','Peter','Natasha','Wanda'}
>>> names
{'Peter', 'Wanda', 'Tony', 'Natasha'}
>>> more_names = {'Steve', 'Peter', 'Carol', 'Wanda'}
>>> more_names
{'Carol', 'Peter', 'Steve', 'Wanda'}
>>> names | more_names
{'Tony', 'Steve', 'Peter', 'Carol', 'Wanda', 'Natasha'}
>>> names
{'Peter', 'Wanda', 'Tony', 'Natasha'}
>>> more_names
{'Carol', 'Peter', 'Steve', 'Wanda'}
>>> names.update(more_names)
>>> names
{'Tony', 'Steve', 'Peter', 'Carol', 'Wanda', 'Natasha'}
```

从上述代码可以看出,两个集合使用"|"运算符得到的运算结果与使用 update() 方法返回结果相同。这两种方法的最大区别是,当使用"|"运算符时,不会更改任何一个集合;但是使用 update() 方法时,会更改使用该方法的集合。因此,在本例中,names 集合更改为了 names|more_names 运算的结果。

```
>>> names = {'Tony','Peter','Natasha','Wanda'}
>>> names
{'Peter', 'Wanda', 'Tony', 'Natasha'}
>>> more_names = {'Steve', 'Peter', 'Carol', 'Wanda'}
>>> more_names
{'Carol', 'Peter', 'Steve', 'Wanda'}
>>> names & more_names
{'Peter', 'Wanda'}
>>> names
```

```
{'Peter', 'Wanda', 'Tony', 'Natasha'}
>>> more_names
{'Carol', 'Peter', 'Steve', 'Wanda'}
>>> names.intersection_update(more_names)
>>> names
{'Peter', 'Wanda'}
```

与 update()方法一样,intersection_update()方法执行"&"操作,但将操作结果赋值给调用 intersection_update()方法的集合。同样地,symmetric_difference_update()方法执行"^"操作,移除当前集合中在另外一个指定集合相同的元素,并将另外一个指定集合中不同的元素插入当前集合中。而 difference_update()方法求取两个集合的差集,移除两个集合都包含的元素①。

本章将介绍的最后一个概念是不可变集合(frozen set),就像元组与列表的关系一样,不可变集合不能更改。示例代码如下:

```
>>> frozen_names = frozenset({'Tony','Peter','Natasha','Wanda'})
>>> frozen_names
frozenset({'Peter', 'Wanda', 'Tony', 'Natasha'})
>>> frozen_names = frozenset(['Tony','Peter','Natasha','Wanda'])
>>> frozen_names
frozenset({'Peter', 'Wanda', 'Tony', 'Natasha'})
>>> frozen_names = frozenset(('Tony','Peter','Natasha','Wanda'))
>>> frozen_names
frozenset({'Peter', 'Wanda', 'Tony', 'Natasha'})
>>> frozen_names = frozenset('Tony')
>>> frozen_names
frozenset({'o', 'n', 'y', 'T'})
>>> dir(frozen_names)
['__and__', '__class__', '__class_getitem__', '__contains__', '__delattr__', '__dir__', '__doc__', '__eq__', '__format__', '__ge__', '__getattribute__', '__gt__', '__hash__', '__init__', '__init_subclass__', '__iter__', '__le__', '__len__', '__lt__', '__ne__', '__new__', '__or__', '__rand__', '__reduce__', '__reduce_ex__', '__repr__', '__ror__', '__rsub__', '__rxor__', '__setattr__', '__sizeof__', '__str__', '__sub__', '__subclasshook__', '__xor__', 'copy', 'difference', 'intersection', 'isdisjoint', 'issubset', 'issuperset', 'symmetric_difference', 'union']
```

上述代码中,我们将集合、列表、元组或字符串使用 frozenset()函数创建集合,并使用 dir()函数查看所创建集合对象可用的方法。可以看出,那些能够更改集合的方法对 frozenset 集合不可用,而用于集合比较的方法仍然可用。

本章小结

本章主要介绍了 Python 中集合的概念及使用方法。

---

① 此处为 symmetric_difference_update 和 difference_update 集合操作,已更正原书中存在的问题。

CHAPTER 11

第 11 章　循环与分支结构

本章将介绍编程方面的循环与分支结构，即 for、if、else 和 while 语句。这些语句是程序设计的关键组成部分，允许对对象进行操作。我们在前面讨论了比较和相等的逻辑运算，这也是 if 语句首先要关注的问题。if 语句的关键是能够返回 True（真）或 False（假）值的语句。考虑以下语句：

```
>>> x = 1
>>> x == 1
True
```

上述代码中，第 1 行语句为 x 赋值 1，第 2 行语句将 x 的值与 1 进行比较结果返回 True。借助第 2 行的比较语句，我们可以引入 if 语句：

```
>>> x == 1
True
>>> if x == 1:
...     x = x + 1
...
>>> x
2
```

此时，if 语句所做的便是测试比较语句的返回值，如果为真，就将变量 x 增加 1。if 语句的本质是根据比较结果执行相应的功能，if 语句测试逻辑运算返回值是否为 True，如果为 True，便可以在该语句中实现一定的功能。通过引入 else 语句，可以扩展上述示例的功能。本质上，if 语句处理结果为 True 的情况，而 else 语句处理结果为 False 的情况。

现在，我们想要给 x 变量赋值 2，然后测试 x 是否等于 1，结果返回 False。接下来再设置一个 if-else 语句，如果 x 的值等于 1，便将 x 增加 1，如果 x 值不等于 1，便将 x 减少 1。示例代码如下：

```
>>> x = 2
>>> x == 1
False
>>> if x == 1:
...     x = x + 1
... else:
...     x = x - 1
...
>>> x
1
```

上述的逻辑非常有用,借助该逻辑我们可以根据变量的取值让程序进行分支选择。然而,if-else 循环只允许根据单个条件的 True 或 False 做出选择。如果需要更多条件的控制,可以在 if-else 语句结构中增加 elif 语句。现在,引入 elif 语句对上述示例进行改进,示例代码如下:

```
>>> x = 2
>>> x == 1
False
>>> x == 2
True
>>> if x == 1:
...     x = x + 1
... elif x == 2:
...     x = x * x
... else:
...     x = x - 1
...
>>> x
4
```

在此,我们将变量 x 设置为 2,然后用设计好的逻辑进行实现。第一个 if 语句测试 x 是否等于 1,结果不能成立,然后使用 elif 语句查看 x 是否等于 2,该条件满足,因此在语句中运行代码计算 x 的平方。在 if-else 分支结构中,可以扩展更多的 elif 语句,以便检查多种情况。需要注意的是,如果满足 if 或 elif 分支上其中一个条件,就将退出 if-else 语句并继续执行后续代码。因此,在考虑使用 if 语句时,需要了解使用它们的上下文。

如果有一组固定的结果要处理,则可使用上面展示的 if-elif 语句。接下来,我们将展示一个使用 if-elif 语句不太恰当的示例,示例代码如下:

```
>>> home_score = 4
>>> away_score = 0
>>> if home_score > away_score:
...     result = "Home win"
... elif home_score < away_score:
...     result = "Away Win"
```

```
... elif home_score == away_score:
...     result = "Draw"
... elif home_score >(away_score + 1):
...     result = "Home win by more than 1 goal"
... elif(home_score + 1) < away_score:
...     result = "Away win by more than 1 goal"
... else:
...     result = "Unknown result"
...
>>> result
'Home win'
```

从表面上看可能会认为这是一段非常好的代码,但事实并非如此。首先要搞明白问题是什么,是想测试哪一方获胜,还是测试至少赢 1 球呢?实际上,在我们测试至少赢 1 球(结果也会返回 True)之前,主场胜利的条件已经满足了。因此,要把两种情况一块考虑,单独使用 if 语句是行不通的,需要使用嵌套的 if 语句。

```
>>> home_score = 4
>>> away_score = 0
>>> if home_score > away_score:
...     if home_score >(away_score + 1):
...         result = "Home win by more than 1 goal"
...     else:
...         result = "Home win"
... elif home_score < away_score:
...     if(home_score + 1) < away_score:
...         result = "Away win by more than 1 goal"
...     else:
...         result = "Away win"
... elif home_score == away_score:
...     result = "Draw"
... else:
...     result = "Unknown result"
...
>>> result
'Home win by more than 1 goal'
```

可以看出,上述程序取得了预期的结果。值得注意的是,if 语句非常强大,但如果没有将所有条件考虑全面,就容易出现错误。接下来将介绍循环的概念。

前面章节介绍了列表(第 7 章)、元组(第 8 章)和字典(第 9 章),它们都是 Python 中的重要容器。我们展示了如何访问和操作它们。然而,人们通常专注于这些对象的单个实例,但实际上往往会有多种数据,每个数据都可能包含在列表、元组或字典中,需要对其进行访问并执行某种操作。为了验证这一说法,让我们创建一些数据:

```
>>> people = []
>>> person = ["Tony", "Stark",48]
```

```
>>> people.append(person)
>>> person = ["Steve","Rodgers",102]
>>> people.append(person)
>>> person = ["Stephen", "Strange",42]
>>> people.append(person)
>>> person = ["Natasha","Romanof",36]
>>> people.append(person)
>>> person = ["Peter","Parker",16]
>>> people.append(person)
>>> people
[['Tony', 'Stark', 48], ['Steve', 'Rodgers', 102], ['Stephen', 'Strange', 42], ['Natasha',
'Romanof', 36], ['Peter', 'Parker', 16]]
```

在此,我们建立了一个由列表组成的列表,现在如果访问第 1 个列表元素中的第 3 个元素,可以按照下述方式进行[①]:

```
>>> people
[['Tony', 'Stark', 48], ['Steve', 'Rodgers', 102], ['Stephen', 'Strange', 42], ['Natasha',
'Romanof', 36], ['Peter', 'Parker', 16]]
>>> people[0][2]
48
```

因此,如果访问外部列表的第 1 个元素,该元素返回一个列表,然后通过索引 2 访问该列表中值为 48 的元素。现在,如果我们想查看每个人的年龄,可以编写以下代码实现:

```
>>> people[0][2]
48
>>> people[1][2]
102
>>> people[2][2]
42
>>> people[3][2]
36
>>> people[4][2]
16
```

以这种方式查看 5 个人的年龄相当乏味,想象一下,如果有成百上千的人呢?为了避免反复编写相同的代码,可以使用循环来访问列表中每个人的年龄。现在,为了让编写的程序更加有趣,让我们来计算列表中所有人的平均年龄。

```
>>> total_age = 0.0
>>> for p in people:
...     age = p[2]
```

[①] 原书本段代码结果有误,进行了更正。

```
...     total_age += age
...
>>> total_age
244.0
>>> average_age = total_age / len(people)
>>> average_age
48.8
```

以上示例主要是介绍循环的概念。上述代码中,首先设置一个变量 total_age 保存总年龄,并将其初始值设置为零,在循环中将每个人的年龄加入到 total_ag 变量中,然后进入 for 循环。对于"for p in people"的语法来说,该语句完成对 people 中所有元素的遍历,并将其赋值给变量 p。变量名 p 是任意取的名字,也可以将循环重写成如下形式:

```
>>> total_age += age
>>> total_age = 0.0
>>> for huge_massive_robot in people:
...     age = huge_massive_robot[2]
...     total_age += age
```

该名称在技术上并不重要,本段代码中将 p 修改为了 huge_massive_robot,尽管命名不是很好,但实现功能相同。关键是,当编写循环时并不必总是使用 p。循环程序执行时,会一个接一个地访问 people 列表中的所有元素,其中每个元素都是一个列表,循环中的缩进代码实现对遍历元素的处理。在循环中所做的是将列表元素的第 3 个值赋给变量 age,然后将该值添加到初始值为零的 total_age 变量中。将该循环代码展开,可以得到如下的代码(循环代码相当于以下内容的简写):

```
>>> age = people[0][2]
>>> total_age += age
>>> age = people[1][2]
>>> total_age += age
>>> age = people[2][2]
>>> total_age += age
>>> age = people[3][2]
>>> total_age += age
>>> age = people[4][2]
>>> total_age += age
>>> total_age
244.0
```

因此,可以看出循环可以大大减少编写的代码量,是一件好事。在循环代码之后,我们用 len() 函数获取列表 people 中的元素个数,然后用总年龄除以该值,就得到了平均年龄。

这是将 for 循环应用于列表的基础,关键是对列表的遍历,要遍历到列表的每个元素。正如我们之前讨论的,列表元素的类型不一定相同,因此在使用循环时需要小心,因为应用于每一个元素的逻辑可能不适用于遍历到的所有元素。这里针对列表所展示的内容也适用于元组，对于元组也可以编写类似的循环，但不能对其进行设置，因为元组不可修改。

基于列表和循环可以实现列表推导式（list comprehension）。现在，如果现在想要创建一个包含人们年龄平方的列表，可以编写如下循环来实现：

```
>>> squared_ages = []
>>> for p in people:
...     squared_age = p[2] * p[2]
...     squared_ages.append(squared_age)
...
>>> squared_ages
[2304, 10404, 1764, 1296, 256]
```

这段代码看起来很好，但是该段代码可以用一行语句来实现，如下所示：

```
>>> squared_ages = [p[2] * p[2] for p in people]
>>> squared_ages
[2304, 10404, 1764, 1296, 256]
```

哪一种实现更好呢？两种方法都做得一样，有些人可能会说使用列表推导式代码更具有 Python 风格，但有人会说它比标准 for 循环更难阅读。使用哪种方法取决于个人喜好和风格。

接下来，我们将研究如何将循环应用于字典，使用方法与列表和元组不同。现在，如果有一个字典的列表，我们可以遍历该列表并访问该字典，就像在上一个示例中访问列表中的列表一样。但是，如果只有一个字典，我们可以按以下方式进行循环操作：

```
>>> person = {"first_name":"Steve", "last_name":"Rodgers", "age":102}
>>> person_list = []
>>> for p in person:
...     person_list.append(p)
...
>>> person_list
['first_name', 'last_name', 'age']
```

上述代码中，循环的行为方式不同于之前列表中的用法。p 是字典的键，而不是字典的键和值，所以在将 p 放入列表中时，会得到字典的键。如果想从 person 字典中获取值，需要按照以下方式进行：

```
>>> person = {"first_name":"Steve", "last_name":"Rodgers", "age":102}
>>> person_list = []
>>> for p in person:
...     value = person[p]
...     person_list.append(value)
```

```
...
>>> person_list
['Steve', 'Rodgers', 102]
```

此处代码与上例代码的唯一区别是在循环中通过字典的键访问对应的值。

接下来，我们看一下怎样使用循环处理字符串。我们可以用与列表相同的循环方式处理字符串，如下所示：

```
>>> name = "Rob Mastrodomenico"
>>> name_list = []
>>> for n in name:
...     name_list.append(n)
...
>>> name_list
['R', 'o', 'b', ' ', 'M', 'a', 's', 't', 'r', 'o', 'd', 'o', 'm', 'e', 'n', 'i', 'c', 'o']
```

因此，当上述循环执行时，我们访问字符串中的每个元素，将其添加到列表中。因此，添加到列表中的是 name 字符串中的每个字母。

本节中介绍的最后一个概念是 while 循环，该循环在某些逻辑为 True 时继续循环。演示示例如下：

```
>>> score = 0
>>> while score < 4:
...     score
...     score += 1
...
0
1
2
3
```

这段程序做了什么呢？首先，设置了一个名称为 score 的变量，并将其初始值设置为 0。接下来，当 score<4 时执行 while 循环中的代码。语法上，需要在循环条件后添加冒号。从输出结果可以看到，当 score<4 时显示 score 的值，一旦条件不满足就退出循环。while 循环与 for 循环很像，然而，for 循环是遍历所有元素，而 while 循环一直要循环到循环条件不满足为止。对于 while 这样的循环，编程时需要特别小心，确保有满足退出的条件，否则会进入死循环！

下面，以一个模拟抽奖的程序为例，将上面讲述的内容包含其中。这里的抽奖游戏是从固定数量的球中随机抽取，给出一组数字。如果恰好选择了这组号码，则表示赢了彩票。下面将展示的示例中生成 6 个普通球和 1 个奖励球。为了随机生成球，使用以下代码：

```
>>> from random import randint
>>> ball = randint(min, max)
```

本例中,球的号码的最小值和最大值分别为 1 和 59,使用 randint()函数可以生成一个 1~59 的随机数。

```
>>> from random import randint
>>> min = 1
>>> max = 59
>>> ball = randint(min, max)
>>> ball
15
```

这里的关键是生成的随机整数并不总是唯一的,可能会得到相同的数字,因此需要有一种方法来确保不会得到相同的数字。在 Python 中有多种方法可确保这一点,我们将介绍两种不同的方法,第一种方法如下:

```
>>> from random import randint
>>> min = 1
>>> max = 59
>>> result_list = []
>>> for i in range(7):
...     ball = randint(min, max)
...     while ball in result_list:
...         ball = randint(min, max)
...     result_list.append(ball)
...
>>> result_list
[11, 57, 3, 10, 38, 41, 42]
```

上述方法中,创建了一个列表来存储想要的结果。然后,通过一个 range 对象实现循环:

```
>>> range(7)
range(0, 7)
```

这样就创建了一个可供循环的 range 对象,因为想要生成 7 个数字,所以 range()函数的参数设置为 7。然后,随机生成一个球,如果生成的球已在列表中,则使用 while 循环重新生成,直至生成一个新的球,然后将其添加到结果列表中。这样就可以得到 7 个随机数来模拟彩票抽奖,也包括奖励球。

第二种方法稍微做了修改,去掉了 range 对象,示例代码如下:

```
>>> from random import randint
>>> min = 1
>>> max = 59
>>> result_list = []
```

```
>>> while len(result_list)< 7:
...     ball = randint(min, max)
...     if ball not in result_list:
...         result_list.append(ball)
...
>>> result_list
[6, 22, 42, 21, 11, 54, 43]
```

上述代码中,while 循环一直执行到 result_list 中包含了所有球的数量为止,这样就可以实现同样功能的同时减少了代码行的数量。

本章小结

本章介绍了 for、if、else 和 while 的概念,展示了如何应用这些概念来解决问题,同时,也将之前学习的内容串联起来。

CHAPTER 12

第 12 章 字 符 串

Python 擅长做很多事情,尤其擅长字符串操作。本章将扩展我们对字符串的了解,展示 Python 在处理字符串方面的强大能力。首先,让我们先看看字符串是什么,示例代码如下:

```
>>> name = 'rob'
>>> name
'rob'
>>> name = "rob"
>>> name
'rob'
>>> name = """rob"""
>>> name
'rob'
>>> name = '''rob'''
>>> name
'rob'
```

在上面的例子中,给出了几种不同的字符串创建方式,其实,创建字符串的方式并不重要,用法都是一样的。这主要围绕单引号和双引号的使用。如果将单引号放在由单引号括起来的字符串中,则会得到以下结果:

```
>>> single_quote_string = 'string with a 'single quote'
  File "<stdin>", line 1
    single_quote_string = 'string with a 'single quote'
                                         ^
SyntaxError: invalid syntax
```

要在由单引号括起来的字符串中包含单引号,需要使用转义序列。一般来说,转义序列是前面带有反斜杠(\)的字符。如果使用转义序列书写上面的字符串,则结果如下:

```
>>> single_quote_string = 'string with a \' single quote'
>>> single_quote_string
"string with a ' single quote"
```

可以用双引号重写上面的例子,如下所示:

```
>>> single_quote_string = "string with a \' single quote"
>>> single_quote_string
"string with a ' single quote"
```

如果使用三重引号,则可以不使用转义序列。使用三重引号重写前面的内容,如下所示:

```
>>> single_quote_string = """string with a ' single quote"""
>>> single_quote_string
"string with a ' single quote"
```

但如果在字符串末尾有双引号,还是会遇到问题。

```
>>> double_quote_string = """string with a double quote on the end"""
    File "<stdin>", line 1
        double_quote_string = """string with a double quote on the end"""
                                                                        ^
SyntaxError: unterminated string literal(detected at line 1)
```

要解决此问题,需要在双引号上使用转义序列,如下所示:

```
>>> double_quote_string = """string with a double quote on the end\""""
>>> double_quote_string
'string with a double quote on the end"'
```

使用三重引号创建字符串的另一个好处是可以在多行上书写字符串。

```
>>> on_two_lines = """A quote on
...                    two lines"""
>>> on_two_lines
'A quote on\n\t\t\t\ttwo lines'
```

上述字符串包含 1 个 \n(行)和 3 个 \t(制表符)。如果尝试将三重引号改为双引号,则会出现以下结果:

```
>>> on_two_lines = "A quote on
    File "<stdin>", line 1
        on_two_lines = "A quote on
                                  ^
SyntaxError: unterminated string literal(detected at line 1)
```

前面已经简要介绍了转义序列,尽管还没有涵盖所有的转义序列(可以自行到网上查找),但出现了一个有趣的问题:如果要在字符串中使用转义序列,则该怎么做。例如,键入\n会在字符串中返回一个回车,但是如果我们想在字符串中使用\n,则可以使用原生字符串(raw string)。示例代码如下:

```
>>> raw_string = r"This has a \n in it"
>>> raw_string
'This has a \\n in it'
>>> not_raw_string = "This has a \\n in it"
>>> not_raw_string
'This has a \\n in it'
```

在这两个例子中,当显示字符串的内容时,会显示一个额外的反斜杠,但当使用print语句打印输出时,额外的反斜杠会消失。如果想再加一个斜杠,就要用三个斜杠。

```
>>> not_raw_string = "This has a \\\n in it"
>>> not_raw_string
'This has a \\\n in it'
```

下面介绍如何获取字符串中所包含的元素。可以用访问列表的方式访问字符串。例如,如果想要获取字符串的第3个元素,可以这样做:

```
>>> name = "Rob Mastrodomenico"
>>> name[2]
'b'
```

需要注意的是,字符串与列表一样是零索引(索引从0开始),因此第3个元素的索引值为2。同样,可以获取第5~8位置上的元素值,如下所示:

```
>>> name = "Rob Mastrodomenico"
>>> name[4:8]
'Mast'
```

字符串同样可以用反向索引,索引值使用负数,如果要获取字符串的最后3个元素,可以按如下操作:

```
>>> name = "Rob Mastrodomenico"
>>> name[-3:]
'ico'
```

或者,我们也可以得到除最后3个元素之外的所有元素。

```
>>> name = "Rob Mastrodomenico"
>>> name[:-3]
'Rob Mastrodomen'
```

我们可以将前面显示的大部分逻辑应用于字符串。因此,如果想知道字符串是否包含某个字符或某个子串,可以通过查看字符是否在字符串中来实现。

```
>>> name = "Rob Mastrodomenico"
>>> "ico" in name
True
>>> "z" in name
False
```

借助代码中的变量可以创建自定义字符串,通常被称作字符串格式化,因此如果想要将变量放入字符串中,只需要使用花括号定义在字符串中插入的位置,然后使用format()方法将变量值作为参数传入到指定的位置。

```
>>> first_name = "Rob"
>>> last_name = "Mastrodomenico"
>>> name = "First name: {}, Last name: {}".format(first_name, last_name)
>>> name
'First name: Rob, Last name: Mastrodomenico'
```

上述示例中的方法似乎没有带来任何好处,因为我们可以像以前那样轻松地定义整个字符串。但是,需要考虑名称发生更改的情况。对名称列表或文件进行循环处理,每次循环时,我们都希望显示名称所对应的值,此时使用上述方法就会非常方便。

我们也可以在花括号中指定变量赋值的位置序号,代码如下:

```
>>> first_name = "Rob"
>>> last_name = "Mastrodomenico"
>>> name = "First name: {1}, Last name: {0}".format(first_name, last_name)
>>> name
'First name: Mastrodomenico, Last name: Rob'
```

上述代码的结果是错的,但你可以从中理解位置序号的作用。我们还可以将每个值定义为一个变量,并将该变量名放在花括号中。

```
>>> first_name = "Rob"
>>> last_name = "Mastrodomenico"
>>> name = "First name: {f}, Last name: {l}".format(f=first_name, l=last_name)
>>> name
'First name: Rob, Last name: Mastrodomenico'
```

下面介绍字符串方法。首先介绍的字符串方法是大小写转换方法。例如,将字符串的所有字母转换为小写字母,然后再转换为大写字母。示例代码如下:

```
>>> name = "Rob Mastrodomenico"
>>> name
'Rob Mastrodomenico'
>>> name.lower()
'rob mastrodomenico'
>>> name.upper()
```

```
'ROB MASTRODOMENICO'
>>> name
'Rob Mastrodomenico'
```

可以看出,使用 lower()方法和 upper()方法创建了字符串的大写和小写的初始版本,但是并不改变字符串本身。这种方法十分有用,当需要检查字符串中的某些字符大小写是否正确时,可借助它将字符串更改为正确的大小写。

接下来介绍 split()方法,它可以用于拆分字符串。

```
>>> name = "Rob Mastrodomenico"
>>> name
'Rob Mastrodomenico'
>>> name.split(" ")
['Rob', 'Mastrodomenico']
>>> first_name, last_name = name.split(" ")
>>> first_name
'Rob'
>>> last_name
'Mastrodomenico'
```

同样,虽然这个例子看起来微不足道,但它非常有用,因为可以在字符串中的任何字符上进行拆分。例如,如果能在 csv 文件中找到以逗号分隔的字符串,就可以拆分出变量对应的值。示例代码如下:

```
>>> match_details = "Manchester United,Arsenal,2,0"
>>> match_details
'Manchester United,Arsenal,2,0'
>>> match_details.split(",")
['Manchester United', 'Arsenal', '2', '0']
>>> home_team, away_team = match_details.split(",")[0:2]
>>> home_team
'Manchester United'
>>> away_team
'Arsenal'
>>> home_goals, away_goals = match_details.split(",")[2:4]
>>> home_goals
'2'
>>> away_goals
'0'
```

因此,从一行字符串中我们就可以得到它包含的所有信息。

字符串的另一个有用方法是 replace()方法,可以执行字符串替换功能。

```
>>> match_details = "Manchester United,Arsenal,2,0"
>>> match_details
'Manchester United,Arsenal,2,0'
```

```
>>> match_details.replace(",",":")
'Manchester United:Arsenal:2:0'
```

上述示例中用冒号替换了所有的逗号,同样也可以像刚才一样更改字符串中的分隔符,该功能非常有用。

还有另一个字符串方法,它表面看起来像一个列表方法,即 join()方法,可实现字符串的拼接。

```
>>> details = ['Manchester United', 'Arsenal', '2', '0']
>>> match_details = ','.join(details)
>>> match_details
'Manchester United,Arsenal,2,0'
```

上述示例对只包含逗号的字符串应用了 join()方法。该示例创建了一个由字符串组成的列表,然后利用 join()方法将列表的字符串元素使用逗号分隔拼接起来。当需要创建以某个常见值分割的列表字符串时该方法非常有用。

最后介绍 len()函数,它是 Python 内置函数,可以应用于字符串。在前面介绍列表时,我们知道 len()函数可以应用于列表以获取其长度,对于字符串也是如此,可以获取字符串的长度。借助 len()函数查看字符串包含多少个字符的示例代码如下:

```
>>> match_details = "Manchester United,Arsenal,2,0"
>>> match_details
'Manchester United,Arsenal,2,0'
>>> len(match_details)
29
```

本章小结

本章深入研究了字符串的概念以及用法,并展示了 Python 在处理已有字符串类型变量或者创建自己的变量时的灵活性以及丰富的功能。

第 13 章 正则表达式

正则表达式是字符串处理的扩展。广义上来说,借助正则表达式可以进行比简单字符串搜索更加高效的高级搜索。

```
>>> text = "Hello in this is string"
>>> 'Hello' in text
True
```

要做到这一点,可以使用 Python 标准库中的正则表达式(regular expression,re)模块,而不是遍历所有的细节,下面通过一些示例进行说明。

第一个示例将查找字符串"Rob Mastrodomenico"中 a~m 的所有字符。为了实现该功能,我们使用 re 模块的 findall()方法。其中,name 是要进行字符串查找的数据源,是要匹配的字符串;[a-m]是要查找的正则表达式匹配模式,表示从字符串中查找所有在 a~m 的字符。由此产生的结果是一个列表,该列表包含出现在字符串中的 a~m 的所有字符。示例代码如下:

```
>>> import re
>>> name = 'Rob Mastrodomenico'
>>> x = re.findall("[a-m]", name)
>>> x
['b', 'a', 'd', 'm', 'e', 'i', 'c']
```

接下来,我们将了解如何在序列对象中找到整数 0~9。有两种方法可以实现,第一种方法是模仿上例中使用的正则表达式;第二种方法是在正则表达式中使用"\d"代表匹配数字查找 0~9 所有的数字。两种方法都返回数字字符串的列表。观察下面展示的示例代码:在第一个 txt 例子中,可以看到我们得到了每个值,但是如果有重复数值呢?在第二个 txt 例子中,可以看到列表中返回了重复的值,因为字符串中有 2 个 3。

```
>>> import re
>>> txt = 'Find all numerical values like 1, 2, 3'
```

```
>>> x = re.findall("\d", txt)
>>> x
['1', '2', '3']
>>> x = re.findall("[0-9]", txt)
>>> x
['1', '2', '3']
>>> txt = 'Find all numerical values like 1, 2, 3, 3'
>>> x = re.findall("[0-9]", txt)
>>> x
['1', '2', '3', '3']
>>> x = re.findall("\d", txt)
>>> x
['1', '2', '3', '3']
```

这种查找字符串中是否存在一个或多个值并给出所有值的方法很有用,但我们可以实现更多功能,并查找特定格式的数据。在下一个示例中,首先以 txt 字符串"hello world"为文本进行查找特定的字符。我们可以将字符串"he..o"传递给 findall()方法,这样做的目的是搜索以"he"开头,紧跟任意两个字符,最后以字符"o"结尾的序列。因为单词"hello"与之匹配,所以运行结果返回包含字符串"hello"的列表。接着,可以通过将 txt 字符串更改为"hello helpo hesoo"进行扩展,可以看到所有这些单词都从 findall()方法中传回。通过使用这样一个不同的例子,可以看到如何将该方法应用于更大的文本,以查看与该序列匹配的所有单词。示例代码如下:

```
>>> import re
>>> txt = "hello world"
>>> x = re.findall("he..o", txt)
>>> x
['hello']
>>> txt = "hello helpo hesoo"
>>> x = re.findall("he..o", txt)
>>> x
['hello', 'helpo', 'hesoo']
```

接下来,看一下怎样搜索字符串的开头。为此,需要在要搜索的字符串前面加"^"符号。在本例中,我们将查找以"start"开头的字符串,得到的结果是一个包含要找单词的列表。因此,在下面示例的第一部分得到了包含字符串"start"的列表,在第二部分得到了一个空列表。

```
>>> import re
>>> txt = 'starts at the end'
>>> x = re.findall("^start", txt)
>>> x
['start']
>>> txt = 'ends at the start'
```

```
>>> x = re.findall("^start", txt)
>>> x
[]
```

同样地，可以通过在要搜索的字符串结尾加"$"符号查看字符串中的最后一个单词。下面示例展示了搜索给定字符串的结尾部分时得到的结果，与前面示例类似，如果该字符串存在就返回一个包含该字符串的列表，如果不存在则返回一个空列表。

```
>>> import re
>>> txt = 'the last word is end'
>>> x = re.findall("end$", txt)
>>> x
['end']
>>> txt = 'the last word is end sometimes'
>>> x = re.findall("end$", txt)
>>> x
[]
```

下面的两个示例着眼于在给定字符串的开始或结束处查找特定的内容。在下一个示例中，我们将查看给定字符串在另一个字符串中的所有实例。该正则表达式是要找到所有的"ai"字符串，其中"ai"后面可以跟0个或多个"x"字符。因此，第一个示例搜索"aix"时，搜索到字符串中有4个"ai"实例。与前面的例子一样，如果我们没有找到任何实例，则返回一个空列表。

```
>>> import re
>>> txt = "The rain in Spain falls mainly in the plain!"
>>> x = re.findall("aix*", txt)
>>> x
['ai', 'ai', 'ai', 'ai']
>>> txt = 'This isnt like the other'
>>> x = re.findall("aix*", txt)
>>> x
[]
```

在上一示例的基础上，可以通过添加符号"+"查找字符串中"ai"的实例，此时，"ai"后面可以跟一个或多个"x"，如下面的示例所示。仍然使用前面的字符串作为待查找字符串，结果返回空的列表，因为要找的字符串中没有"aix"。

```
>>> import re
>>> txt = "The rain in Spain falls mainly in the plain!"
>>> x = re.findall("aix+", txt)
>>> x
[]
```

如果要查找指定数量的字符，可以使用包含要查找字符数量的花括号。因此，在下面的示例中，我们希望在字符串中查找"moo"，因此传入的搜索字符串可以是"mo{2}"或

"moo",两种情况都返回了一个包含搜索字符串的列表。

```
>>> import re
>>> txt = 'The cow said moo'
>>> x = re.findall("mo{2}", txt)
>>> x
['moo']
>>> x = re.findall("moo", txt)
>>> x
['moo']
```

如果想查找两个值中的其中一个,可以在两个查找字符串之间使用"|"符号来实现。在下面的示例中,我们尝试查找字符串中的"avengers"或"heroes"。语句①表示在被搜索字符串中查找"avengers"或"heroes",因为查找字符串中有一个"Avengers",但其以大写 A 开头,所以只有一个"heroes"准确匹配。语句②表示在被搜索字符串中查找"Avengers"或"heroes",因为使用了大写字母 A 的"Avengers"作为查找字符串的一部分,所以得到了包含这两个字符串的列表。语句③中,由于在被搜索字符串中有待查找单词"Avengers"的多个实例,最终将按照找到它们的顺序列出相应的实例。

```
>>> import re
>>> txt = "The Avengers are earths mightiest heroes"
>>> x = re.findall("avengers|heroes", txt) ①
>>> x
['heroes']
>>> x = re.findall("Avengers|heroes", txt) ②
>>> x
['Avengers', 'heroes']
>>> txt = "The Avengers are earths mightiest heroes go Avengers"
>>> x = re.findall("Avengers|heroes", txt) ③
>>> x
['Avengers', 'heroes', 'Avengers']
```

在 re 正则表达式中还可以使用下面这样的特殊序列,它返回给定字符串中的空格。

```
>>> txt = "Is there whitespace"
>>> x = re.findall("\s", txt)
>>> x
[' ', ' ']
```

在正则表达式中,还可以使用其他特殊表达式序列,如表 13-1 所示。

表 13-1　特殊表达式序列

匹 配 符	说　　明	示　　例
\A	只在字符串的开头进行匹配	"\AIt"
\b	匹配位于开头或结尾的字符串	"\bain" r"ain\b"

续表

匹配符	说明	示例
\B	匹配不位于开头或结尾的字符串(前面的 r 用于确保字符串是 raw 字符串)	r"\Bain" r"ain\B"
\d	匹配任意十进制数(0~9 的数字)	"\d"
\D	返回字符串中不包含数字的匹配项	"\D"
\s	返回字符串中包含空白字符的匹配项	"\s"
\S	返回字符串中不包含空白字符的匹配项	"\S"
\w	返回字符串中包含任何数字和字符(0~9 的数字和 a~z、A~Z、下划线"_"字符)的匹配项	"\w"
\W	返回字符串中不包含任何数字和字符的匹配项	"\W"
\Z	返回在字符串末尾进行字符匹配的匹配项	

下面，使用 re 的 split()方法进行示例说明：

```
>>> import re
>>> txt = "The rain in Spain"
>>> x = re.split("\s", txt)
>>> x
['The', 'rain', 'in', 'Spain']
```

在此基础上，我们可以使用 maxsplit 参数指定要执行拆分的次数。下面的示例将该值设置为 1、2 和 3。在每个 re 示例中，我们都会增加拆分的数量，因此将 maxsplit 设为 1 将得到一个拆分了 1 次的列表，列表中有 2 个元素。随着 maxsplit 参数的增加，可以得到越来越多的分割。

```
>>> import re
>>> txt = "The rain in Spain"
>>> x = re.split("\s", txt, maxsplit = 1)
>>> x
['The', 'rain in Spain']
>>> x = re.split("\s", txt, maxsplit = 2)
>>> x
['The', 'rain', 'in Spain']
>>> x = re.split("\s", txt, maxsplit = 3)
>>> x
['The', 'rain', 'in', 'Spain']
```

下一个方法是 sub()方法，类似于字符串替换。下面的示例将空格替换为 9。

```
>>> import re
>>> txt = "The rain in Spain"
>>> x = re.sub("\s", "9", txt)
>>> x
'The9rain9in9Spain'
```

与前面的 split() 示例一样,还有一个额外的参数 count 可以使用,下面 3 个 sub() 方法示例中 count 分别取值为 1、2 和 3。当 count 取值为 1 时,字符串仅将第一个空格替换为 9;当 count 取值为 2 时,字符串将前两个空格替换为 9,以此类推。

```
>>> import re
>>> txt = "The rain in Spain"
>>> x = re.sub("\s", "9", txt, 1)
>>> x
'The9rain in Spain'
>>> x = re.sub("\s", "9", txt, 2)
>>> x
'The9rain9in Spain'
>>> x = re.sub("\s", "9", txt, 3)
>>> x
'The9rain9in9Spain'
```

最后一个例子是在 search() 方法的结果中使用 span() 方法,可以返回匹配位置组成的元组。本例中,search() 方法找到了两个 ai 实例,调用 span() 方法返回的结果为 (5,7),其中 5 是第一个 ai 出现的位置,7 是从第一次找到 ai 的位置之后第 2 个 ai 所在的位置。

```
>>> import re
>>> txt = "The rain in Spain"
>>> x = re.search("ai", txt)
>>> x.span()
(5, 7)
```

本章小结

本章的示例很好地演示了正则表达式模块的使用,当我们对字符串进行复杂搜索时,正则表达式便能展现出其强大的功能。

CHAPTER 14

第 14 章 文件操作

到目前为止,我们已经介绍了 Python 的主要功能,并通过简单的示例进行了讲解。事实上,我们希望处理某种类型的数据,需要将其导入 Python 进行处理。现在,数据的来源有很多,我们可以从数据库、Web API 等各种来源获取数据,但常见获取数据的方式还是通过传统的文件方式。因此,本章将利用学过的知识进行文件数据的读写操作。

Python 可以很容易地从标准库中读取文件,可以根据文件所在的位置创建文件流。假定 test.csv 文件中存储了由逗号分隔的数据,已知该文件其存储的路径为/Path/to/file/test.csv,下面演示如何打开该文件,示例代码如下:

```
>>> file_name = "/Path/to/file/test.csv"
>>> f = open(file_name,"r")
>>> f
<_io.TextIOWrapper name = '/Path/to/file/test.csv' mode = 'r' encoding = 'cp936'>
```

上述代码中定义了一个包含文件名的字符串 file_name,通过调用 open()函数读取该文件,其中参数为 file_name 和"r"(read,读模式)。变量 f 为 open()函数的返回值,是一个文件流(stream)。要从文件中读取数据,只需运行如下代码:

```
>>> data = f.read()
```

上述操作返回的是单索引列表数据,这种数据用处不大。在文本文件中,不同的行由换行符"\n"进行分隔,可以应用 split()方法将读取的内容以换行符为界拆分为列表的新元素。下面借助字符串进行演示,示例代码如下:

```
>>> names = "steve\ntony\nbruce\n"
>>> names
'steve\ntony\nbruce\n'
>>> names.split("\n")
['steve', 'tony', 'bruce', '']
```

因此,我们可以将数据转换成可读的格式。文件中的行也是由换行符分隔,分隔方式与上述字符串相同。可按如下操作从文件中读取数据：

```
>>> data = f.read().split("\n")
```

对于逗号分隔的文件，文件中的行由逗号分隔的元素组成。所以，为了得到每个元素，我们需要根据逗号对每一行进行再次拆分。为此，需要执行以下操作：

```
>>> names = "steve,rodgers\ntony,stark\nbruce,banner\n"
>>> names
'steve,rodgers\ntony,stark\nbruce,banner\n'
>>> names = names.split("\n")
>>> names
['steve,rodgers', 'tony,stark', 'bruce,banner', '']
>>> for n in names:
...     row = n.split(",")
...     row
...
['steve', 'rodgers']
['tony', 'stark']
['bruce', 'banner']
['']
```

最初，我们将字符串拆分成一个包含 3 个元素的列表。然后在列表上执行循环操作，对于列表中的每个元素用逗号分隔，重新创建一个包含名字和姓氏的列表。如果不使用 names 列表循环，也可以借助列表推导式在一行中完成该项操作。示例代码如下：

```
>>> names = "steve,rodgers\ntony,stark\nbruce,banner\n"
>>> names
'steve,rodgers\ntony,stark\nbruce,banner\n'
>>> names = names.split("\n")
>>> names
['steve,rodgers', 'tony,stark', 'bruce,banner', '']
>>> names = [n.split(",") for n in names]
>>> names
[['steve', 'rodgers'], ['tony', 'stark'], ['bruce', 'banner'], ['']]
```

由上可见，无论是利用列表循环还是利用列表推导式，两者得出结果的末尾都得到了一个空列表，主要原因是在最后一行返回换行符时，对于该换行符进行拆分就会得到一个空字符串。因此，对于以换行符(\n)分隔的任何文件都需要考虑空字符串。借助前面介绍的 pop() 方法可以解决这一问题：

```
>>> names
[['steve', 'rodgers'], ['tony', 'stark'], ['bruce', 'banner'], ['']]
>>> names.pop()
['']
>>> names
[['steve', 'rodgers'], ['tony', 'stark'], ['bruce', 'banner']]
```

现在，我们已经可以读取文件了。接下来要介绍的是如何写入文件，其工作方式与读取文件非常相似，同样需要首先定义一个文件名，然后打开一个流执行文件写入。

```
>>> file_name = "output.csv"
>>> f = open(file_name, "w")
>>> f
<_io.TextIOWrapper name = 'output.csv' mode = 'w' encoding = 'cp936'>
```

上述代码将在终端窗口以写入模式打开一个流，将某些内容在物理上写入文件，为此，需要定义需要写入文件的内容。

```
>>> out_str = "something to go in the file\n"
>>> f.write(out_str)
28
>>> f.close()
```

现在，我们创建一个要写入文件的字符串，然后使用文件流形式将该字符串写入文件。在第一个读取文件的示例中遗漏了一件事，即忘记了关闭文件流。上述示例中，在最后一行中使用 close() 方法实现了文件的关闭。当退出 Python 或结束编写的程序时，Python 通常会进行资源回收，但最好能在代码中包含这一点。

接下来，我们将介绍如何在文件中追加内容。该项操作与写入文件类似，但当以写入模式打开文件时会覆盖任何现有的同名文件。使用追加模式时，现有文件将得以保留，然后在其末尾添加任何想要的内容。追加模式与之前写入文件的方法类似，只是在打开文件时使用 "a"（append，追加模式）选项。如果要在 output.csv 文件中追加内容，需要编写以下内容：

```
>>> file_name = "output.csv"
>>> f = open(file_name, "a")
>>> f
<_io.TextIOWrapper name = 'output.csv' mode = 'a' encoding = 'cp936'>
```

将上述示例进行扩展以便读写更大的文件，通过从 sklearn 模块中导入数据集[①]实现，示例代码如下：

```
>>> from sklearn.datasets import load_boston
>>> boston = load_boston()
```

接下来，加载一个字典对象，其中包含一个数据集和要处理的相关细节。此处需要从这个字典中获取数据和 feature_name 键，并写入 output.csv 文件。示例代码如下：

---

① 在执行该段代码时，系统提示 "load_boston" 在 1.0 中已弃用，并将在 1.2 中删除，提示波士顿房价数据集存在道德问题，只是出于教育研究目的才可以使用。本书后面的部分数据集在运行代码时也会出现该提示，但可以运行出结果。

```
>>> feature_names = boston['feature_names'][::2]
>>> list(feature_names)
['CRIM', 'INDUS', 'NOX', 'AGE', 'RAD', 'PTRATIO', 'LSTAT']
```

为了增加难度,我们将进行间隔取值,并且不包括最后两个值,示例代码如下:

```
>>> headers = list(feature_names)[::2][:-2]
>>> headers
['CRIM', 'NOX']
```

上述示例结果提供了要写入输出文件的数据,接下来我们将打开文件并将 headers 数据写入文件。

```
>>> file_name = "boston_output.csv"
>>> fo = open(file_name, "w")
>>> fo.write(','.join(headers) + '\n')
9
```

从上述示例可以看到,输出的 9 表示向文件中写入了 9 个字符[①]。接下来要做的是将引用 headers 数据写入文件。

```
>>> boston_data = boston['data']
>>> for bd in boston_data:
...     row_dict = dict(zip(feature_names, bd))
...     val_list = []
...     for h in headers:
...         val = row_dict[h]
...         val_list.append(str(val))
...     out_str = ','.join(val_list)
...     fo.write(out_str + '\n')
```

上述代码中,首先将 data 分配给 boston_data,然后对 boston_data 进行循环。随后,将 data 中的每个元素与 feature_names 中的数据进行 zip 操作以创建一个字典。这样做的目的是选择要写入文件的相关值。为此,我们循环遍历 headers 列表,使用 headers 的键访问字典值,并将这些值添加到列表中。最后使用 join() 方法将这些值附加到一个列表中并按行写入文件。

```
>>> fo.close()
```

最后需要将文件关闭,这是一种好的习惯。但从技术上讲,如果我们不这样做,Python 也会执行该操作。

接下来,我们可以读入文件并使用以下代码遍历内容。

---

① 9 表示写入文件的字符数,原书 23 有误,实际写入了 9 个字符。

```
>>> file_name = "boston_output.csv"
>>> f = open(file_name, "r")
>>> data = f.read().split('\n')
>>> data.pop()
''
>>> for d in data:
...     print(d)
...
```

这就是关于文件的读取、写入和追加操作的全部内容。需要注意的是,本节所展示的内容仅适用于单页数据文件(single sheet data file)。

14.1 Excel 文件

包含数据表的电子表格文件是 Python 文件操作中常见的文件类型,它可以是 xls 或 xlsx 文件的形式。幸运的是,Python 提供了一个名为 openpyxl 的库,它允许将数据写入 excel 文件或者从 excel 文件中将数据读出,正如下面所演示的一样[①]。

```
>>> from sklearn.datasets import load_boston
>>> boston = load_boston()
>>> feature_names = boston['feature_names'][::2]
>>> list(feature_names)
['CRIM', 'INDUS', 'NOX', 'AGE', 'RAD', 'PTRATIO', 'LSTAT']
>>> headers = list(feature_names)[::2][:-2]
>>> headers
['CRIM', 'NOX']
```

接下来,我们希望将该数据读入一个 excel 数据表:

```
>>> from openpyxl import Workbook
>>> wb = Workbook()
>>> sheet1 = wb.create_sheet('boston_data', 0)
```

在这里要做的首先是导入相关的包(openpyxl),然后创建一个工作簿。对于该工作簿,创建一个要写入的工作表,取名为 boston_data,然后将其插入电子表格的位置 0,成为第一个电子表。

```
>>> i = 1
>>> for h in headers:
...     sheet1.cell(1, i, h)
```

[①] 下述示例代码中用到的波士顿房价数据集存在一些问题,在 scikit-learn 1.0 中已经弃用,将在 1.2 中删除,目前在 Python3.10 中执行 load_boston() 会产生警告。可以参考本例用法,借助 Pandas 和 NumPy 获取住房原始数据集进行分析。

```
...     i += 1
...
<Cell 'boston_data'.A1>
<Cell 'boston_data'.B1>
```

上述代码是将 headers 列表内容写入工作表,并将值插入工作表的第 1 行,因此代码中计数器 i 设置为 1,从第 1 列开始,然后递增,从而将后续值插入相关列。此处使用了 cell() 方法,传入的参数分别是行(row)、列(column)和值(value),其中 row 固定为 1。

```
>>> j = 2
>>> boston_data = boston['data'][0:5]
>>> for bd in boston_data:
...     k = 1
...     row_dict = dict(zip(feature_names, bd))
...     for h in headers:
...         val = row_dict[h]
...         sheet1.cell(j, k, val)
...         k += 1
...     j =+ 1
...
<Cell 'boston_data'.A2>
<Cell 'boston_data'.B2>
<Cell 'boston_data'.A1>
<Cell 'boston_data'.B1>
<Cell 'boston_data'.A1>
<Cell 'boston_data'.B1>
<Cell 'boston_data'.A1>
<Cell 'boston_data'.B1>
<Cell 'boston_data'.A1>
<Cell 'boston_data'.B1>
```

接下来,我们希望将数据的前五行写入文件,因此使用了相同的 cell() 方法,但是现在需要增加行和列来处理多行操作。为此,行计数器设置在外循环,内循环进行列处理,主要是因为对于每一行都需要将列重置。因为需要返回第 1 列,所以每次完成一行的写入时,k 值都需要更改为 1。

```
>>> wb.save('boston.xlsx')
```

最后是保存数据,在工作簿上使用 save() 方法并输入要保存的文件名。
读取数据的过程相对简单。

```
>>> from openpyxl import load_workbook
>>> wb = load_workbook('boston.xlsx')
>>> wb
<openpyxl.workbook.workbook.Workbook object at 0x00000196FFEC7E80>
```

通过 worksheets 方法,可以查看有哪些工作表。

```
>>> wb.worksheets
[<Worksheet "boston_data">, <Worksheet "Sheet">]
>>> sheet = wb['boston_data']
>>> sheet
<Worksheet "boston_data">
```

然后，可以使用字典表示法（dictionary notation）访问特定的工作表，将工作表名称视为键，使用行和列索引获取值。

```
>>> sheet[1][0].value
'CRIM'
>>> sheet[1][1].value
'NOX'
```

在此需要注意，由于列的索引是从零开始的，尽管在代码中写入文件指明写入第1列，但可以通过value属性获取特定值。

14.2 JSON 文件

作为一种非常流行的数据类型，JSON（JavaScript Object Notation）是一种轻量级数据交换格式，被业界广泛使用。但JSON有什么实际意义呢？它实际上是一种用于数据存储的文本格式，具有直观性和易用性，易于机器解析和生成，读写数据更加方便。

对于Python用户来说，JSON看起来是列表和字典的混合体，因为JSON中既有像字典中的键值对数据，也有像列表一样存储的数据。下面以之前使用过的示例为例，创建数据的JSON表示。

```
>>> from sklearn.datasets import load_boston
>>> boston = load_boston()
>>> feature_names = boston['feature_names'][::2]
>>> list(feature_names)
['CRIM', 'INDUS', 'NOX', 'AGE', 'RAD', 'PTRATIO', 'LSTAT']
>>> headers = list(feature_names)[::2][:-2]
>>> boston_data = boston['data'][0:5]
>>> boston_data
```

上述代码主要内容与前面示例相同，不同之处在于只选择了数据的前五个元素，这样可以允许以JSON表示方式显示数据。

```
>>> json_list = []
>>> for bd in boston_data:
...     row_dict = dict(zip(feature_names, bd))
...     val_dict = {}
...     for h in headers:
```

```
...            val = row_dict[h]
...            val_dict[h] = val
...        json_list.append(val_dict)
...
>>> print(json_list)
[{'CRIM': 0.00632, 'NOX': 2.31}, {'CRIM': 0.02731, 'NOX': 7.07}, {'CRIM': 0.02729, 'NOX':
7.07}, {'CRIM': 0.03237, 'NOX': 2.18}, {'CRIM': 0.06905, 'NOX': 2.18}]
```

上述代码将数据转换为 JSON 的格式。如前所述，可以通过字典和列表的组合实现。首先创建了一个列表，将数据的每一行放入其中。列表的行可以是一个字典，其键值对由 headers 的两个特征名组成。最终得到的是一个由字典构成的列表 json_list，下面希望将其导出为 JSON 格式。

```
>>> import json
>>> file_name = "boston.json"
>>> with open(file_name, "w") as write_file:
...     json.dump(json_list, write_file, indent = 4)
...
```

上述代码中，为了得到 JSON 输出，使用了 JSON 包及其 dump()方法，将 json_list 和 write_file(一个打开的文件)作为输入参数。

```
[
    {
        "CRIM": 0.00632,
        "NOX": 2.31
    },
    {
        "CRIM": 0.02731,
        "NOX": 7.07
    },
    {
        "CRIM": 0.02729,
        "NOX": 7.07
    },
    {
        "CRIM": 0.03237,
        "NOX": 2.18
    },
    {
        "CRIM": 0.06905,
        "NOX": 2.18
    }
]
```

下面介绍如何使用 Python 读取 JSON 文件，可借助 JSON 包轻松实现。示例代码如下：

```
>>> import json
>>> file_name = "boston_output.json"
>>> with open(file_name, "r") as read_file:
... data = json.load(read_file)
...
>>> data
[{'CRIM': 0.00632, 'NOX': 2.31}, {'CRIM': 0.02731, 'NOX': 7.07},
{'CRIM': 0.02729, 'NOX': 7.07}, {'CRIM': 0.03237, 'NOX': 2.18},
{'CRIM': 0.06905, 'NOX': 2.18}]
>>> type(data)
<class 'list'>
```

与本章开头介绍的读取文件的方法类似，读取 JSON 文件时，只使用具有打开文件模式读取的加载方法，最后将文件中的值分配给类型为列表的数据对象。

14.3 XML 文件

可扩展标记语言（extensible markup language，XML）是一种存储数据的方式，与 JSON 非常相似，可以轻松地读取和写入数据，方便计算机进行解析和生成数据。与 JSON 不同，XML 没有与 Python 数据类型的自然联系，因此本节将更多地介绍 XML 的类型及其工作方式。让我们用下面的例子进行解释。

```xml
<?xml version = "1.0"?>
<catalog>
    <book id = "bk101">
        <author> Rob, Mastrodomenico </author>
        <title> The Python book </title>
        <genre> Computer </genre>
        <price> Whatever </price>
        <publish_date> Whenever </publish_date>
        <description> Stuff to help you master Python </description>
    </book>
    <book id = "bk102">
        <author> Rob, Mastrodomenico </author>
        <title> The Python book 2 </title>
        <genre> Computer </genre>
        <price> More than the last one </price>
        <publish_date> Maybe never </publish_date>
        <description> Its like the first one but better </description>
    </book>
</catalog>
```

现在让我们来解构上面的示例。

```xml
<?xml version = "1.0"?>
```

第一行是 XML 声明,可以简单书写如下:

```
<?xml?>
```

如果在 XML 文件中使用了一些特定的编码,可以在该行中用 encoding 指定编码:

```
<?xml version = "1.0" encoding = "utf-8"?>
```

接下来,我们看如下语句:

```
<catalog>
</catalog>
```

两个 catalog 之间内容是 XML 的根元素(root element),是 XML 内容开始和结束的标志。标签名称的使用是任意的,在本例中,它只反映了我们所拥有的数据。您可能会注意到,在第二个 catalog 前面有一个"/"符号,这是开始和结束元素常见的形式。

接下来,添加了一个深一层级的内容,如下所示:

```
<book id = "bk101">
</book>
```

在这里,我们定义了一个以<book>开始和</book>结束的 book 标签。与根级别不同,我们通过添加 id="bk101"将数据附加到这个级别。为了在 book 中添加更具体的数据,可以这样做:

```
<book id = "bk101">
    <author> Rob, Mastrodomenico </author>
    <title> The Python book </title>
    <genre> Computer </genre>
    <price> Whatever </price>
    <publish_date> Whenever </publish_date>
    <description> Stuff to help you master Python </description>
</book>
```

在 book 级别下,我们添加了 author、title、genre、price、publish_date 和 description 等标签。如前所述,每个标签的定义都有一个开头和结尾。

如果要添加另一个 book 标签,可按如下操作:

```
<book id = "bk102">
    <author> Rob, Mastrodomenico </author>
    <title> The Python book 2 </title>
    <genre> Computer </genre>
    <price> More than the last one </price>
    <publish_date> Maybe never </publish_date>
    <description> Its like the first one but better </description>
</book>
```

与之前一样，在第一个 book 标签下面可以创建另一个 book 标签，依据每个 book 标签中的 id 值来对 book 标签进行区分。

我们在这里展示的是如何使用 XML 构建有趣的数据结构。下面来看一下如何创建和解析 XML 对象。为此，我们使用 lxml 库，它是一个 Python 库，允许用户使用 C 库的 libxml2 和 libxslt。这些都是非常快速的 XML 处理库，可以通过 Python 轻松访问。

正如本章前面所做的，我们将使用相同的示例展示如何从中创建 XML。

```
>>> from sklearn.datasets import load_boston
>>> boston = load_boston()
>>> feature_names = boston['feature_names'][::2]
>>> list(feature_names)
['CRIM', 'ZN', 'INDUS', 'CHAS', 'NOX', 'RM', 'AGE', 'DIS',
'RAD', 'TAX', 'PTRATIO', 'B', 'LSTAT']
>>> headers = list(feature_names)[::2][:-2]
>>> boston_data = boston['data'][0:5]
```

将数据写入 XML 的完整代码如下：

```
>>> from lxml import etree
>>> root = etree.Element("root")
>>> for bd in boston_data:
...     row_dict = dict(zip(feature_names, bd))
...     row = etree.SubElement(root, "row")
...     for h in headers:
...         child = etree.SubElement(row, h)
...         val = row_dict[h]
...         child.text = str(val)
>>> et = etree.ElementTree(root)
>>> et.write('boston.xml', pretty_print = True)
```

我们首先导入 lxml 库的 etree 模块，接着创建 XML 文档名为 root 的标签元素。

```
>>> from lxml import etree
>>> root = etree.Element("root")
```

接下来，采用与之前类似的方式循环获取数据，并将数据放入 XML 文件中。

```
>>> for bd in boston_data:
...     row_dict = dict(zip(feature_names, bd))
...     row = etree.SubElement(root, "row")
...     for h in headers:
...         child = etree.SubElement(row, h)
...         val = row_dict[h]
...         child.text = str(val)
```

数据处理的循环机制与之前的示例没有什么不同。在循环中，同样创建了 row_dict 变量，并针对 headers 进行循环以获取数据值，但不同的是如何设置 xml 以及将数据写入哪里。对于 boston_data 的每次迭代，使用 SubElement() 方法在 root 创建新的一行，取名为 row，并将 root 指定为父节点。然后，对于从 headers 循环中获得的每个值，创建另一个子节点，此时将 row 作为父节点，名称为从 headers 列表获得的值。最后，通过将子节点的 text 属赋值的形式进行设置得到了所需的数据格式。

```
>>> et = etree.ElementTree(root)
>>> et.write('boston.xml', pretty_print = True)
```

最后一部分是将数据写入文件，通过在 ElementTree() 方法中传入 root 参数，并使用 write() 方法实现。请注意，此处将 pretty_print 设置为 True，这将生成以下文件：

```
<root>
    <row>
        <CRIM>0.00632</CRIM>
        <NOX>2.31</NOX>
    </row>
    <row>
        <CRIM>0.02731</CRIM>
        <NOX>7.07</NOX>
    </row>
    <row>
        <CRIM>0.02729</CRIM>
        <NOX>7.07</NOX>
    </row>
    <row>
        <CRIM>0.03237</CRIM>
        <NOX>2.18</NOX>
    </row>
    <row>
        <CRIM>0.06905</CRIM>
        <NOX>2.18</NOX>
    </row>
</root>
```

下面演示如何在 Python 中使用 lxml 库读取 XML 文件。

```
>>> from lxml import objectify
>>> xml = objectify.parse(open('boston.xml'))
>>> root = xml.getroot()
>>> children = root.getchildren()
>>> print(children)
[< Element row at 0x1f4604899c0 >, < Element row at 0x1f460489840 >, < Element row at 0x1f460489240 >, < Element row at 0x1f460498080 >, < Element row at 0x1f460499680 >]
>>> for c in children:
...     print(c['CRIM'])
...     print(c['NOX'])
...
0.00632
```

```
2.31
0.02731
7.07
0.02729
7.07
0.03237
2.18
0.06905
2.18
```

上述示例通过从 lxml 库导入 objectify 模块实现 XML 文档的读取。下面分条进行解析。

```
>>> xml = objectify.parse(open(file_name))
```

首先，使用 objectify 的 parse() 方法读取 XML 文件对其进行解析，结果返回一个 XML 对象，可以使用它来解析文档信息。接下来，可以使用以下方法获取文档的根节点：

```
>>> root = xml.getroot()
```

得到根节点之后，我们希望获取表示下一级的子节点，即 row 节点。

```
>>> children = root.getchildren()
```

现在，由于 children 对象只是一个列表，可以遍历 children 对象访问 row 节点中的值。可以利用以下方法获得并打印输出：

```
>>> for c in children:
...     print(c['CRIM'])
...     print(c['NOX'])
```

上述输出的是我们在数据集中创建的值。

本章小结

本章介绍了与文件有关的一些重要概念，以及如何使用 Python 进行文件读写。文中介绍了许多不同的文件类型，并给出了针对性的操作示例。本书后续章节还会介绍其他文件读写的方法，但当我们采用高层次方法处理数据时，这些底层的数据处理方法显得非常重要，为后续进行高级数据处理打下基础。

CHAPTER 15

第 15 章 函数与类

本章将介绍函数和类的概念,以及如何将这些概念引入 Python 代码中。到目前为止,本书中的所有内容都以代码块的形式呈现。这些代码块可能只是演示特定功能的实现,也可能是演示如何执行某项复杂任务的部分功能。在编程入门阶段,借助这种方式编写代码是完全可以接受的。然而,如果想实现代码的分组与重用,就要用到函数和类的概念。

首先来看一下函数的概念。函数可以实现代码的重用,如果有可重复的代码,借助函数可以快速实现这类代码的使用。函数还可以带有参数,能根据传递给它的变量决定函数的行为。对于第 11 章给出的模拟抽奖示例,下面演示如何使用函数来实现。模拟抽奖示例代码如下:

```
>>> from random import randint
>>> min = 1
>>> max = 59
>>> result_list = []
>>> while len(result_list)< 7:
...     ball = randint(min, max)
...     if ball not in result_list:
...         result_list.append(ball)
...
>>> result_list
[52, 58, 22, 32, 46, 56, 14]
```

现在,我们可以把上述模拟抽奖示例代码转化成一个名为 lottery 的函数,如下所示:

```
>>> from random import randint
>>> def lottery():
...     min = 1
...     max = 59
...     result_list = []
...     while len(result_list)< 7:
```

```
...         ball = randint(min, max)
...         if ball not in result_list:
...             result_list.append(ball)
...     return result_list
```

上述代码中定义了一个名为 lottery 的函数,函数的定义使用 def 命令,后面紧跟函数名和小括号,小括号里面是要传递给函数的参数。此例中,括号内没有任何内容,表示没有向函数传递参数。请注意,在小括号后面需要跟一个冒号,这与 if、else、for 和 while 语句使用方法相同。从 def 语句开始,后面的代码与前面示例中的代码完全相同,但需要将代码从定义函数的位置缩进一级,另外,在最后增加了 return 语句,将 return 与 result_list 变量一起使用,目的是将函数的运行结果变量 result_list 返回给函数调用者。在本例中,我们返回的是生成的彩票号码列表。要运行上述函数,只需执行以下操作:

```
>>> results = lottery()
>>> results
[57, 9, 7, 22, 23, 5, 32]
```

考虑到函数实现的功能,代码中采用 randint() 函数产生随机数。如果希望得到 1~49 的数,而不是 1~59 的数,可以在代码中将 59 修改为 49。但实际上,如果只是修改变量 max 和 min 的值来改变产生随机数的范围,这样做的意义不大。通过对 lottery() 函数进行重写,可以更加方便地实现这一功能,如下所示:

```
>>> from random import randint
>>> def lottery(min, max):
...     result_list = []
...     while len(result_list)< 7:
...         ball = randint(min, max)
...         if ball not in result_list:
...             result_list.append(ball)
...     return result_list
...
```

上述代码中,我们将 min 和 max 作为参数传递到 lottery() 函数中。尽管这些值是最小值和最大值,但我们可以随意调用它们。这两个参数变量可以取名为 min 和 max,也可以取名为 x 和 y,不影响实现的功能,只关系到代码中引用问题。在 lottery() 函数中,min 和 max 只在 randint() 函数中进行了使用,以返回规定范围的随机数。下面程序演示了 lottery() 函数的使用方法:

```
>>> results = lottery(1, 49)
>>> results
[9, 46, 29, 22, 40, 27, 32]
```

在上一个示例中,我们在函数中传入了要在函数中使用的最小值和最大值,但是有时候可能希望它们有一个默认值,如最初示例中的 1 和 59。可以通过给函数参数赋默认

值的方式来实现,具体如下:

```
>>> from random import randint
>>> def lottery(min = 1, max = 59):
...     result_list = []
...     while len(result_list)< 7:
...         ball = randint(min, max)
...         if ball not in result_list:
...             result_list.append(ball)
...     return result_list
...
```

这里,通过使用赋值的形式,将 min 默认为 1,将 max 默认为 59。这样,调用函数就可以更加灵活,如下所示:

```
>>> lottery()
[26, 11, 46, 22, 44, 15, 34]
>>> lottery(1,49)
[9, 26, 4, 39, 33, 46, 20]
>>> lottery(max = 49)
[23, 8, 15, 3, 38, 37, 9]
```

从上面的结果可以看出,如果不给 lottery()函数传入任何参数,则 min 和 max 默认取值为 1 和 59。如果给 lottery()函数传入 2 个参数,则第 1 个参数赋值给了 min,第 2 个参数赋值给了 max。如果只给 lottery()函数的 max 参数传入了数值,则 min 参数使用了默认值。本例说明,在使用函数的时候使用参数可以增加函数的灵活性。

如果仔细研究前面编写的 lottery()函数示例,你会发现,还可以对其修改使其更加灵活。到目前为止,我们使用的代码允许在抽签中准确生成 7 个球,但有时可能希望得到更多或更少的球。为此,我们可以向函数定义中再传递一个变量 draw_length 来实现这一功能,如下所示:

```
>>> from random import randint
>>> def lottery(min = 1, max = 59, draw_length = 7):
...     result_list = []
...     while len(result_list)< draw_length:
...         ball = randint(min, max)
...         if ball not in result_list:
...             result_list.append(ball)
...     return result_list
...
```

运用上述示例,对其进行调用,可以得到如下结果:

```
>>> lottery()
[22, 14, 37, 3, 11, 7, 24]
>>> lottery(1,49,6)
```

```
[22, 8, 23, 45, 37, 17]
>>> lottery(max = 49,draw_length = 6)
[29, 45, 44, 34, 7, 23]
```

上述 lottery() 的工作方式与前面看到的完全相同，但每次调用它时，都需要准确传入所需的参数，如果传入的参数类型不同会发生什么呢？

```
>>> lottery('1','49','6')
Traceback(most recent call last):
    File "<stdin>", line 1, in <module>
    File "<stdin>", line 3, in lottery
TypeError: '<' not supported between instances of 'int' and 'str'
```

此时，函数返回了一个错误，因为它不支持传入的数据类型。因此，在函数中判断是否可以处理传入的参数是有意义的。为此，我们需要将传入的参数值 min 和 max 指定为整数，因为 randint() 函数生成整数；类似地，抽签数量也只能是整数，因为它与列表的长度有关。可以重写函数如下：

```
>>> def lottery(min = 1, max = 59, draw_length = 7):
...     if type(min)!= int:
...         print('min must be int')
...         return None
...     if type(max)!= int:
...         print('max must be int')
...         return None
...     if type(draw_length)!= int:
...         print('draw_length must be int')
...         return None
...     result_list = []
...     while len(result_list)< draw_length:
...         ball = randint(min, max)
...         if ball not in result_list:
...             result_list.append(ball)
...     return result_list
...
```

上述代码中，使用内置函数 type() 检查每个变量传入的值是否为整数。请注意，这是逐个完成的，以便当类型不对时可以发送提示信息，说明错误发生的原因和问题所在。要查看效果，只需运行以下命令：

```
>>> lottery('1','49','6')
min must be int
```

从上面的运行结果可以看到，函数返回的消息"min must be int"，提示传递给参数 min 的值必须是 int 类型。但是，本例中传递给 max 和 draw-length 的数据也有问题，因为传入的是字符串，而要求是整数。为此，我们可以使用 if 和 else 语句的组合来扩展上面的逻辑，以确定哪些变量传入了无效的数据类型。

```python
>>> def lottery(min = 1, max = 59, draw_length = 7):
...     min_val = True
...     max_val = True
...     draw_length_val = True
...
...     if type(min) != int:
...         min_val = False
...     if type(max) != int:
...         max_val = False
...     if type(draw_length) != int:
...         draw_length_val = False
...
...     if min_val is False:
...         if max_val is False:
...             if draw_length_val is False:
...                 print('min, max, draw_length need to be integer')
...                 return
...             else:
...                 print('min and max need to be integer')
...                 return
...         else:
...             if draw_length_val is False:
...                 print('min and draw_length need to be integer')
...                 return
...             else:
...                 print('min needs to be integer')
...                 return
...     else:
...         if max_val is False:
...             if draw_length_val is False:
...                 print('max and draw_length need to be integer')
...                 return
...             else:
...                 print('max need to be integer')
...                 return
...         else:
...             if draw_length_val is False:
...                 print('draw_length needs to be integer')
...                 return
...             else:
...                 pass
...
...     result_list = []
...     while len(result_list)< draw_length:
...         ball = randint(min, max)
...         if ball not in result_list:
...             result_list.append(ball)
...     return result_list
...
```

上面示例中，我们针对每个输入参数都设置了一个布尔变量，默认取值为 True。如果传入参数的类型不是整数，将对应布尔变量设置为 False。程序使用 if-else 语句判断参数类型的正确性,并将类型不正确的参数进行打印输出。请注意,当输入参数类型不完全正确时函数不会返回任何内容,返回类型为 None。

```
>>> lottery(1,'2',3)
max need to be integer
>>> lottery('1',2,3)
min needs to be integer
>>> lottery('1','2',3)
min and max need to be integer
>>> lottery('1',2,'3')
min and draw_length need to be integer
>>> lottery('1','2','3')
min, max, draw_length need to be integer
>>> res = lottery(1,'2','3')
max and draw_length need to be integer
>>> res is None
True
```

在介绍了函数之后,下面将讨论 Python 中的类(class)。类是一种功能非常强大的对象,可以将许多函数和变量组合在一起。类中定义的函数和变量一般称为方法和属性。我们将通过创建一个简单的类来演示这一点:

```
>>> class MyClass:
...     x = 10
...
>>> mc = MyClass()
>>> mc.x
10
```

上述示例使用 class 关键字创建了一个名为 MyClass 的类。在该类中,定义了一个值为 10 的变量 x。然后,创建了 MyClass 的一个实例 mc,并使用点语法(dot syntax)访问类中的变量 x。下面,我们对之前创建的模拟抽奖示例函数进行扩展。

```
>>> class Lottery:
...     def __init__(self, min = 1, max = 59, draw_length = 7):
...         self.min = min
...         self.max = max
...         self.draw_length = draw_length
...
...     def lottery(self):
...         min_val = True
...         max_val = True
...         draw_length_val = True
...
...         if type(self.min) != int:
...             min_val = False
...         if type(self.max) != int:
...             max_val = False
...         if type(self.draw_length) != int:
...             draw_length_val = False
```

```
...            if min_val is False:
...                if max_val is False:
...                    if draw_length_val is False:
...                        print('min, max, draw_length need to be integer')
...                        return
...                    else:
...                        print('min and max need to be integer')
...                        return
...                else:
...                    if draw_length_val is False:
...                        print('min and draw_length need to be integer')
...                        return
...                    else:
...                        print('min needs to be integer')
...                        return
...            else:
...                if max_val is False:
...                    if draw_length_val is False:
...                        print('max and draw_length need to be integer')
...                        return
...                    else:
...                        print('max need to be integer')
...                        return
...                else:
...                    if draw_length_val is False:
...                        print('draw_length needs to be integer')
...                        return
...                    else:
...                        pass
...
...            result_list = [ ]
...            while len(result_list) < self.draw_length:
...                ball = randint(self.min, self.max)
...                if ball not in result_list:
...                    result_list.append(ball)
...            return result_list
...
```

在这段代码中,我们使用之前的方式定义了一个名为 Lottery 的类。在 Lottery 类代码中,使用__init__创建了一个 init 方法。当定义类的对象时,会调用该初始化程序,以便传入可以在类中使用的参数。请注意,可以在类中定义的函数 lottery() 中使用默认值,但这些默认值是通过 self 参数借助点语法(self-dot syntax)传入类中,该语法允许传入的变量值成为该类的一部分(通常称作类的属性),并允许在类中的任何位置使用。以下示例创建了类的不同对象实例:

```
>>> l = Lottery()
>>> l.lottery()
```

```
>>> l = Lottery(1,49,6)
>>> l.lottery()
>>> l = Lottery(draw_length = 8)
>>> l.lottery()
```

您可能会认为这与函数没有太大的区别。实际上，借助类来实现功能可以让代码更改更加容易。对于前面使用函数处理类型不正确变量的情形，下面用类的方式进行修改。

```
>>> from random import randint
>>> class Lottery:
...     def __init__(self, min = 1, max = 59, draw_length = 7):
...         self.min = min
...         self.max = max
...         self.draw_length = draw_length
...         self.valid_data = True
...         if type(self.min) != int:
...             print('min value needs to be of type int')
...             self.valid_data = False
...         if type(self.max) != int:
...             print('max value needs to be of type int')
...             self.valid_data = False
...         if type(self.draw_length) != int:
...             print('draw_length value needs to be of type int')
...             self.valid_data = False
...
...     def lottery(self):
...         if self.valid_data:
...             min_val = True
...             max_val = True
...             draw_length_val = True
...             result_list = []
...             while len(result_list)< self.draw_length:
...                 ball = randint(self.min, self.max)
...                 if ball not in result_list:
...                     result_list.append(ball)
...             return result_list
...
```

上述代码所做的是将检查类型的复杂逻辑从lottery()函数转移到__init__()初始化函数中，这样做便于打印需要更改的内容，并设置有效的数据属性。只有当输入数据有效性返回True时，才能运行lottery()函数。借助类，我们在代码中有了更大的灵活性，可以利用属性和设置逻辑来影响类中其他方法的功能。因此，正确使用类可以使程序的功能更加强大。

在前面的示例中，我们展示了如何编写函数和类，与本书的其余部分一样，都是在交互式shell中完成了程序的演示工作。不过，利用文件方法编写程序更加实用，也很容易实现，只需要将完全相同的代码写入任何空白文件，并用后缀.py保存即可。

虽然理论上可以使用任何编辑器编写代码，但最好使用集成开发环境（IDE）进行代码编写。安装完 Anaconda 之后，会同时安装一款十分有用的 Python IDE 软件 Spyder。有许多免费的 Python 集成开发环境，你可以选择一款适合自己的，并逐步习惯使用文件编写代码。现在，我们将演示如何在 Spyder 中进行开发。Spyder 启动后，便可以看到一个类似于图 15.1 所示的界面。

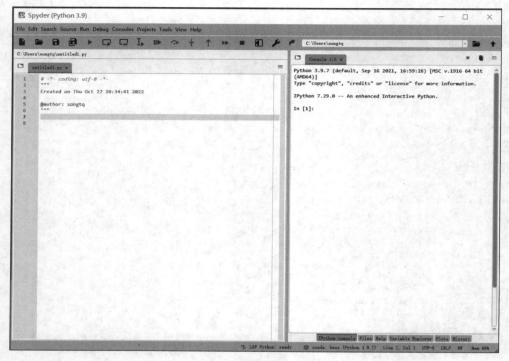

图 15.1 Spyder 集成开发环境

Spyder 开发环境中，最重要的两个窗口是编辑器（Editor）和控制台（Console）。在编辑器窗口可以像本书一样键入代码，并可以将代码保存到物理文件中。为什么要这样做呢？实际上，系统允许将 Python 命令、函数、类或者多种代码组合保存成文件。代码保存为文件之后，可以非常简单地使用 Spyder 之类的集成开发环境重新运行之前编写的代码。因此，通过单击 run 按钮即可执行文件中的所有代码。Spyder 集成开发环境还提供了一个控制台，借助该控制台可以测试命令，也可以访问程序脚本中的变量。下面，举一个非常简单的例子说明这一点，如图 15.2 所示。

在此，我们编写了一个非常简单的脚本程序，取名为 test.py，程序中定义了一个值为 1.0 的变量 x，然后打印出来。通过单击工具栏中的运行文件箭头可以运行这个脚本，该脚本显示在控制台的第 1 行，runfile 参数中显示了运行的文件名，并且程序的运行结果 1.0 也打印到屏幕上。在控制台中也可以访问程序中定义的 x 变量，例如在控制台第 2

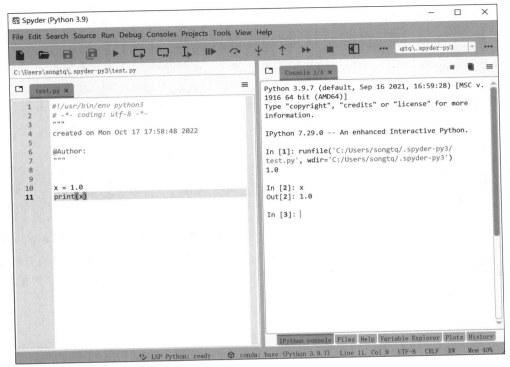

图 15.2　在 Spyder 中运行程序

行输入 x 便会显示变量 x 的值。

将程序代码存入文件是非常有必要的，借助文件可以实现程序的存档、代码的版本控制和代码共享。下面通过例子演示该项功能，使用在本章开头定义的 lottery() 函数，并将其存入名为 lottery.py 的文件中，接下来可以通过将其导入 Python 来使用它。在前面的章节中介绍了如何导入包，该方法同样适用于导入自己编写的 Python 文件。导入自己开发的程序文件时，我们需要了解 Python 是如何进行文件导入的。为此，我们导入 sys 包，并通过 sys.path 查看系统路径列表。

```
>>> import sys
>>> sys.path
['', 'D:\\Program Files\\Python\\Python310\\python310.zip', 'D:\\Program Files\\Python\\Python310\\DLLs', 'D:\\Program Files\\Python\\Python310\\lib', 'D:\\Program Files\\Python\\Python310', 'D:\\Program Files\\Python\\Python310\\lib\\site-packages']
```

上述列表包含了 Python 可能搜索的位置，所以如果想导入 lottery.py 文件，就要将其放入 sys.path 显示列表中的某个路径目录中。请注意，第一个条目是当前程序所在的位置。如果需要，我们还可以将新的路径添加到此列表中。将 lottery.py 文件存入指定目录之后，通过运行以下代码即可导入 lottery.py 文件的内容。

```
>>> from lottery import *
```

导入之后,就可以访问 lottery.py 文件中的所有内容,此时便可以通过调用 lottery() 函数实现抽奖功能。另外,还可以使用前面用于导入包的其他方法导入。允许导入自己编写的代码可以让编程人员灵活地组织代码,对于关键代码可提前编写,在需要时进行共享,以此来降低代码的重复性。我们可以将类导入另一个文件,然后从该文件中运行代码。如果在与 lottery.py 相同的目录下创建一个名为 import_lotting.py 的文件,可以按如下方式运行 Lottery 类:

```
from lottery import *
l = Lottery()
l.lottery()
```

在 Python 中,与其他文件共享代码就这么简单。这样,如果将可共享代码中的函数或类导入自己的代码中,与其他文件共享将变得非常容易。在 Spyder 中,同时拥有控制台和编辑器是很重要的,因为有时可能希望以交互方式了解需要做什么。但总的来说,将代码存入文件进行程序开发会更有效率。

第 16 章 Pandas

前面已经讨论了 Python 标准库(standard library)和包(package)的概念,本章将讨论第三方库 Pandas,它与 Python 生态系统非常相关,主要用于数据分析和数据处理,具有广泛的用途。因此,深入理解并掌握 Pandas 的概念和用法对 Python 程序员至关重要。本章将从 Pandas 的基础知识开始,逐步对 Pandas 相关的先进技术展开介绍。在开始介绍 Pandas 之前,首先简要介绍 NumPy 数组(array),它与列表和字典一样,都是应该理解并掌握的概念。

16.1 NumPy 数组

NumPy(Numerical Python)是 Anaconda 发行版的一部分,是 Python 科学库中的一个关键组件。NumPy 执行速度非常快,是 Python 中其他很多包的基础。本书将只讨论 NumPy 数组,如果读者对 Python 中的关于机器学习的库感兴趣,应更深入地学习 NumPy。

可以按照如下方式导入 NumPy:

```
>>> import numpy as np
```

上述代码中为什么用 np? 这是文档中使用 NumPy 别名的标准约定,可不必使用该约定,但一般都会遵守这一约定。本章不会介绍 NumPy 的所有内容,而是只介绍其中几个概念,首先将介绍的是 NumPy 数组。NumPy 中的数组很像 Python 中的列表。如果要创建一个 0~10 的整数数组,可以按如下方式进行:

```
>>> number_array = np.array([0, 1, 2, 3, 4, 5, 6, 7, 8, 9, 10])
>>> number_array
array([ 0, 1, 2, 3, 4, 5, 6, 7, 8, 9, 10])
```

上述代码看起来像是往 array()方法中传递了一个列表,这是创建数组的通常做法。

我们也可以按照如下方式定义相同的数组：

```
>>> number_list = [0, 1, 2, 3, 4, 5, 6, 7, 8, 9, 10]
>>> number_array = np.array(number_list)
>>> number_array
array([ 0, 1, 2, 3, 4, 5, 6, 7, 8, 9, 10])
```

上述数组看起来像一个列表并且能用列表来创建它，但它与列表有什么不同呢？在前面讲解列表操作的时候发现，如果对列表使用常见的数学运算符要么不起作用，要么以意想不到的方式起作用。现在我们将再次讨论这些问题，并将结果与在 NumPy 数组上操作的结果进行比较。首先看一下加法：

```
>>> number_list + number_list
[0, 1, 2, 3, 4, 5, 6, 7, 8, 9, 10, 0, 1, 2, 3, 4, 5, 6, 7, 8, 9, 10]
>>> number_array + number_array
array([ 0, 2, 4, 6, 8, 10, 12, 14, 16, 18, 20])
```

可以看到，对于列表，就像之前看到的一样，两个列表串联在一起；但如果使用数组，则是将两个数组的元素相加，并返回一个新数组。接下来看一下当使用减法时会得到什么结果。

```
>>> number_list - number_list
Traceback(most recent call last):
  File "<stdin>", line 1, in <module>
TypeError: unsupported operand type(s) for - : 'list' and 'list'
>>> number_array - number_array
array([0, 0, 0, 0, 0, 0, 0, 0, 0, 0, 0])
```

从运行结果可以看到，当按两个列表处理时，结果会按"错误"处理，因为程序不知道如何在两个列表上进行该操作。但如果是两个数组，则会得到预期结果，即从第一个数组中减去第二个数组中对应的元素。如果分别对两个数组和两个列表进行乘法操作时，又会得到什么结果呢？示例代码如下：

```
>>> number_list * number_list
Traceback(most recent call last):
  File "<stdin>", line 1, in <module>
TypeError: can't multiply sequence by non-int of type 'list'
>>> number_array * number_array
array([ 0, 1, 4, 9, 16, 25, 36, 49, 64, 81, 100])
```

可以看到，对于两个列表执行乘法操作同样不起作用，而对于两个数组则进行了对应的元素相乘的操作。现在对列表和数组分别进行除法操作，如下所示：

```
>>> number_list / number_list
Traceback(most recent call last):
  File "<stdin>", line 1, in <module>
```

```
TypeError: unsupported operand type(s) for /: 'list' and 'list'
>>> number_array / number_array
<stdin>:1: RuntimeWarning: invalid value encountered in divide
array([nan, 1., 1., 1., 1., 1., 1., 1., 1., 1.])
```

毫无意外，除法操作对于列表不起任何作用，但对于数组，在执行两个数值中对应元素相除操作时，除了第一个元素得到了零除法警告外，其余的都进行了对应元素的除法操作，使用第一个数组中的元素除以第二个数组的对应元素。数组为我们提供了一种更好、更快的数学运算操作方法，但这些数学操作却不能用列表方法实现。在目前已经讨论过的 Python 示例中，可以利用只有一列的列表达到某些效果。因此，对于上面 3 个（减法、乘法和除法）利用列表不能达到效果的示例可以改写如下：

```
>>> number_list = [0, 1, 2, 3, 4, 5, 6, 7, 8, 9, 10]
>>> subtraction_list = [nl - nl for nl in number_list]
>>> subtraction_list
[0, 0, 0, 0, 0, 0, 0, 0, 0, 0, 0]
>>> multiplication_list = [nl * nl for nl in number_list]
>>> multiplication_list
[0, 1, 4, 9, 16, 25, 36, 49, 64, 81, 100]
>>> division_list = [nl / nl for nl in number_list]
Traceback(most recent call last):
    File "<stdin>", line 1, in <module>
    File "<stdin>", line 1, in <listcomp>
ZeroDivisionError: division by zero
```

从结果可以看到，借助列表推导式使列表也可以完成减法和乘法操作，但由于除数为零，利用列表仍然无法完成除法操作。值得注意的是，现在使用的两个列表是相同的，因而很容易进行重写。但如果是两个不同的列表，我们就无法使用列表推导式，同时使用循环进行重写也变得更加困难。但在 NumPy 数组中，利用随机选择功能创建两个随机数组就能完成这种操作，下面演示这种操作：

```
>>> import numpy as np
>>> np.random.choice(10,10)
array([8, 9, 7, 0, 6, 1, 2, 9, 2, 3])
```

该示例中生成了一个长度为 10、范围为 0～9 的随机数数组。但如果将此示例扩展到 100 万个随机数，并生成两个数组，并将这两个数组对应的元素相乘，示例代码如下所示：

```
>>> x = np.random.choice(100, 1000000)
>>> x
array([16, 46, 56, ..., 56, 99, 57])
>>> y = np.random.choice(100, 1000000)
>>> y
array([16, 27, 11, ..., 19, 70, 17])
```

```
>>> result = x * y
>>> result
array([ 256, 1242, 616, ..., 1064, 6930, 969])
```

在该示例中立即完成100万次乘法,但如果使用循环,则需要等待较长的时间才能获得结果。这说明 NumPy 数组的功能非常强大,幸运的是我们可以像访问列表中的元素一样访问 NumPy 数组中的元素。为了获取上面示例中产生数组中的数据,可以参见下面的示例:

```
>>> result[10]
2970
>>> result[10:20]
array([2970, 760, 1080, 4550, 4420, 4743, 4032, 3569, 3120, 288])
>>> result[:-1]
array([ 810, 2775, 198, ..., 2491, 40, 1120])
>>> result[-3:-1]
array([ 40, 1120])
```

可以看出,访问数组中元素的方式与访问列表中元素的方式是一样的。

至此,已经介绍了与数组相关的概念,下面开始讨论 Pandas 中的 Series 序列数据。

16.2 Series

可以按如下方式导入 Pandas:

```
>>> import pandas as pd
```

与使用 NumPy 一样,在导入 Pandas 时使用别名 pd(Pandas 库的惯用别名)。

首先讨论 Series(序列)[①]的概念。下面通过一个例子来演示:

```
>>> import pandas as pd
>>> point_dict = {"Bulgaria": 45, "Romania": 43, "Israel": 30,
... "Denmark": 42}
>>> point_series = pd.Series(point_dict)
>>> point_series
Bulgaria    45
Romania     43
Israel      30
Denmark     42
dtype: int64
```

---

① Series 是 Pandas 库中的一维数据结构,类似于带有索引的一维数组,本书统一将 Series 译作序列,有些地方是指 Series 对象。

我们创建了一个以国家名称为键、以该国公民年龄中位数为值的字典(来源：sourceworldomometers.info)，并传递给 Series() 方法，得到了序列对象 point_series，并可按如下方式访问该序列中的元素：

```
>>> point_series[0]
45
>>> point_series[1:3]
Romania    43
Israel     30
dtype: int64
>>> point_series[-1]
42
>>> point_series[:-1]
Bulgaria   45
Romania    43
Israel     30
dtype: int64
>>> point_series[[1,3]]
Romania    43
Denmark    42
dtype: int64
```

从上述代码可以看到，可以像访问列表一样，使用所需值的位置访问序列中第一个元素，也可以使用冒号分隔的位置值以及前面提到的负索引获得序列中的元素。另一种方式是通过向序列中传递元素位置的列表来访问序列的元素。例如，当想获得序列中的第一个和第三个元素，则列表中的值就为 0 和 2。

在上述程序中，通过传入索引的方式访问了新创建序列的字典的值，但字典的键呢？它们在序列中有何用途？下面展示将序列当作字典一样进行访问，示例代码如下：

```
>>> point_series.index
Index(['Bulgaria', 'Romania', 'Israel', 'Denmark'], dtype='object')
>>> point_series['Bulgaria']
45
```

可以看出，序列具有 index 属性，它就是字典的键，我们可以使用前面提到的访问字典的方法访问它的值。在早期介绍字典的时候，我们就发现如果试图从字典中访问不存在的键会产生异常，对于序列来说也会产生同样的问题。

```
>>> point_series["England"]
Traceback(most recent call last):
    File "D:\Program Files\Python\Python310\lib\site-packages\pandas\core\indexes\base.py", line 3803, in get_loc
        return self._engine.get_loc(casted_key)
    File "pandas\_libs\index.pyx", line 138, in pandas._libs.index.IndexEngine.get_loc
    File "pandas\_libs\index.pyx", line 165, in pandas._libs.index.IndexEngine.get_loc
```

```
    File "pandas\_libs\hashtable_class_helper.pxi", line 5745, in pandas._libs.hashtable.
PyObjectHashTable.get_item
    File "pandas\_libs\hashtable_class_helper.pxi", line 5753, in pandas._libs.hashtable.
PyObjectHashTable.get_item
KeyError: 'England'

The above exception was the direct cause of the following exception:

Traceback(most recent call last):
    File "<stdin>", line 1, in <module>
    File "D:\Program Files\Python\Python310\lib\site-packages\pandas\core\series.py",
line 981, in __getitem__
        return self._get_value(key)
    File "D:\Program Files\Python\Python310\lib\site-packages\pandas\core\series.py",
line 1089, in _get_value
        loc = self.index.get_loc(label)
    File "D:\Program Files\Python\Python310\lib\site-packages\pandas\core\indexes\
base.py", line 3805, in get_loc
        raise KeyError(key) from err
KeyError: 'England'
>>> point_series.get("England")
```

从上述代码中可以看到,如果访问的键不存在会产生异常,但是也可以看到采用 get() 方法访问 England 所对应的值并没有产生异常,只是返回了 None。

前面已经展示如何访问序列的元素,现在将展示怎样对序列中的元素进行操作。由于序列的提出是基于 NumPy 中数组的概念,因此 NumPy 中的大多数操作也能用于序列。下面创建一个由随机数组成的序列,并展示如何对其进行操作。

```
>>> point_series.get("England")
>>> import numpy as np
>>> np.random.rand(10)
array([0.74013347, 0.25906332, 0.09396589, 0.98791956, 0.90755617,
       0.92967773, 0.76401406, 0.0037732 , 0.18013109, 0.385283  ])
```

采用 NumPy 中的随机(random)方法生成了一个长度为 10、范围为 0~1 的随机数数组,这样便于给序列赋值[①]。

```
>>> import numpy as np
>>> import pandas as pd
>>> random_series = pd.Series(np.random.rand(10))
>>> random_series
0    0.005198
1    0.994883
```

[①] 本处程序执行结果与执行环境有关,为保证运行代码的可靠性,此处列出的结果是采用译者运行的结果,与原书有所不同,后面有些代码的执行结果也具有类似的情况。

```
2    0.679281
3    0.228509
4    0.693374
5    0.082673
6    0.338798
7    0.891500
8    0.532482
9    0.515487
dtype: float64
```

可以用与 NumPy 数组基本相同的方式对该序列进行操作。示例代码如下：

```
>>> import numpy as np
>>> import pandas as pd
>>> random_series_one = pd.Series(np.random.rand(10))
>>> random_series_two = pd.Series(np.random.rand(10))
>>> random_series_one + random_series_two
0    0.318389
1    1.366748
2    0.922710
3    0.614358
4    0.855618
5    1.390782
6    1.095437
7    0.959066
8    0.716339
9    0.913439
dtype: float64
>>> random_series_one - random_series_two
0    0.069607
1   -0.233949
2    0.651076
3   -0.031410
4   -0.267058
5   -0.436646
6    0.164248
7   -0.755803
8    0.609889
9    0.759199
dtype: float64
>>> random_series_one / random_series_two
0    1.559582
1    0.707691
2    5.793775
3    0.902721
4    0.524247
5    0.522119
6    1.352770
7    0.118530
```

```
8    12.458709
9    10.844330
dtype: float64
>>> random_series_one * random_series_two
0    0.024132
1    0.453317
2    0.106873
3    0.094112
4    0.165191
5    0.435904
6    0.293251
7    0.087142
8    0.035294
9    0.064497
dtype: float64
```

上述这些操作看起来与数组的操作是相同的,其最主要的不同是可以对序列进行拼接。示例代码如下:

```
>>> random_series_one[1:] * random_series_two[:-1]
0         NaN
1    0.453317
2    0.106873
3    0.094112
4    0.165191
5    0.435904
6    0.293251
7    0.087142
8    0.035294
9         NaN
dtype: float64
```

我们看到,可以通过索引对序列的元素进行乘法操作。对于序列中两个没有索引的元素,结果则显示为"NaN"。

前面使用字典对序列进行了定义,也可以使用列表或数组定义序列,如下所示:

```
>>> pd.Series([1,2,3,4,5,6])
0    1
1    2
2    3
3    4
4    5
5    6
dtype: int64
>>> pd.Series(np.array([1,2,3,4,5,6]))
0    1
1    2
2    3
```

```
3    4
4    5
5    6
dtype: int32
```

正如上述代码所示，Pandas 会自动为序列添加索引。但是如果想采用特定的索引，可以按如下方式定义：

```
>>> pd.Series([1, 2, 3, 4, 5, 6], index = ["a", "b", "c", "d", "e", "f"])
a    1
b    2
c    3
d    4
e    5
f    6
dtype: int64
```

在此，我们将一个可选列表传递给 index 变量，并将其定义为序列的索引。值得注意的是，索引列表的长度必须与要生成序列的列表或数组的长度相匹配。

16.3　DataFrame

学习过 Series（Pandas 库中的一维数据结构）之后，再来看一下 DataFrame（Pandas 库中的二维数据结构）。DataFrame 可以说是 Pandas 中最受欢迎、也最常用的数据类型，其本质是一种以行和列的格式来对数据进行存储的数据结构，因此有的用户将其当成电子表格或者将其看作数据库表格。

首先讲解如何创建 DataFrame。与序列类似，可以有多种方法创建 DataFrame。

```
>>> countries = ["United Kingdom","France","Germany","Spain","Italy"]
>>> median_age = [40,42,46,45,47]
>>> country_dict = {"name": countries, "median_age": median_age}
>>> country_df = pd.DataFrame(country_dict)
>>> country_df
             name  median_age
0  United Kingdom          40
1          France          42
2         Germany          46
3           Spain          45
4           Italy          47
```

在上面的程序中首先设置了两个列表，一个包含国家名，另一个是公民年龄中位数，并将国家名和公民年龄中位数分别作为键和值存入字典中。然后将这个字典传递到 Pandas 库的 DataFrame() 方法中，得到一个 DataFrame 对象，它的列名分别为 name 和 median_age。从返回结果可以看到，索引被自动定义为 0～4，并与每个列表中的元素相对应。

```
>>> countries = pd.Series( ["United Kingdom","France",
... "Germany","Spain","Italy"])
>>> median_age = pd.Series([40,42,46,45,47])
>>> country_dict = {"name": countries, "median_age": median_age}
>>> country_df = pd.DataFrame(country_dict)
>>> country_df
          name     median_age
0  United Kingdom         40
1          France         42
2         Germany         46
3           Spain         45
4           Italy         47
```

同样地，我们也可以使用序列的字典创建 DataFrame 对象，即重新将序列分配给字典并将其传递到 DataFrame()方法中。如果使用 NumPy 数组，也会得到相同的结果。

接下来使用元组列表创建 DataFrame，其中的数据是国家名称、年龄中位数和人口密度。

```
>>> data = [("United Kingdom", 40, 281),("France", 42, 119),
...("Italy", 46, 206)]
>>> data_df = pd.DataFrame(data)
>>> data_df
               0    1    2
0  United Kingdom  40  281
1          France  42  119
2           Italy  46  206
```

这里，我们创建了一个元组列表，并将列表传递到 DataFrame()方法中，它会返回一个 3×3 的 DataFrame 对象。与之前不同的是，这里不仅自动分配了索引值，而且还自动分配了列名，虽然这些列名不太有用（稍后将会演示如何设置这两个值）。这同样适用于列表的列表、序列的列表、数组的列表，同样也适用于字典的列表，只是表达方式略有不同。

```
>>> data = [{"country": "United Kingdom", "density": 281, "median_age":40},
...{"country": "France", "density": 119, "median_age": 42},
...{"country":"Italy", "density":206, "median_age": 46}]
>>> data_df = pd.DataFrame(data)
>>> data_df
          country  density  median_age
0  United Kingdom      281          40
1          France      119          42
2           Italy      206          46
```

从上述代码可以看出，当传入字典的列表时得到了相同的 DataFrame，但生成的 DataFrame 对象从字典中获得了列的名称。从表面上看，这种方法的工作原理与列表的

列表相同,但是当我们改变某些键时,则会得到某些不同的结果。示例代码如下:

```
>>> data = [{"country": "United Kingdom", "density": 281, "median_age":40},
... {"country": "France", "density": 119, "median_age": 42},
... {"country":"Italy", "density":206, "median": 46}]
>>> data_df = pd.DataFrame(data)
>>> data_df
        country  density  median_age  median
0  United Kingdom      281        40.0     NaN
1          France      119        42.0     NaN
2           Italy      206         NaN    46.0
```

从上述代码可以看出,由于每个字典中的键并非完全相同,Pandas 用"NaN"填充缺失的值[①]。接下来,我们将研究如何访问 DataFrame 的元素。

```
>>> data = [{"country": "United Kingdom", "density": 281, "median_age":40},
... {"country": "France", "density": 119, "median_age": 42},
... {"country":"Italy", "density":206, "median_age": 46}]
>>> data_df = pd.DataFrame(data)
>>> data_df
        country  density  median_age
0  United Kingdom      281          40
1          France      119          42
2           Italy      206          46
>>> data_df['country']
0    United Kingdom
1            France
2             Italy
Name: country, dtype: object
>>> data_df["country"][0]
'United Kingdom'
>>> data_df["country"][0:2]
0    United Kingdom
1            France
Name: country, dtype: object
```

上述示例包含了很多内容。首先,利用字典的列表创建了一个 DataFrame 对象。然后,通过将列的名称作为键传给 DataFrame 获得了名称为 country 的列的所有元素。接着,通过添加所需值的索引来访问该列第一个元素。这里有一个重要的特征,即检索的不是第一个元素,而是检索指定列中索引值为 0 的值。最后,以常用列表方式选择 country 列中索引值为 0 和 1 的两个元素。下面再次检索特定的索引行:

```
>>> data_df["country"][-1]
Traceback(most recent call last):
```

---

① NaN(Not a Number,非数)是计算机科学中数值数据类型的一类值,表示未定义或不可表示的值。NaN 适用于表示一个本来要返回的操作数却未返回数值的情况。

```
        File "D:\Program Files\Python\Python310\lib\site-packages\pandas\core\indexes\
range.py", line 391, in get_loc
            return self._range.index(new_key)
ValueError: -1 is not in range

The above exception was the direct cause of the following exception:

Traceback(most recent call last):
    File "<stdin>", line 1, in <module>
    File "D:\Program Files\Python\Python310\lib\site-packages\pandas\core\series.py",
line 981, in __getitem__
        return self._get_value(key)
    File "D:\Program Files\Python\Python310\lib\site-packages\pandas\core\series.py",
line 1089, in _get_value
        loc = self.index.get_loc(label)
    File "D:\Program Files\Python\Python310\lib\site-packages\pandas\core\indexes\
range.py", line 393, in get_loc
        raise KeyError(key) from err
KeyError: -1
```

正如在列表中所做的一样，这里检索的是 country 列的最后一个值，但由于 DataFrame 的索引中没有值为 -1 的索引，从而引发异常。因此，我们不能像处理列表那样处理 DataFrame，需要理解其索引规则。对于任意的 DataFrame 对象，我们可以获取其索引和列，如下所示：

```
>>> data_df.index
RangeIndex(start=0, stop=3, step=1)
>>> data_df.columns
Index(['country', 'density', 'median_age'], dtype='object')
```

由上述程序可以看出，DataFrame 对象索引的 start 为 0，stop 为 3，step 为 1，并以列表的形式给出了第一列的名称。我们可以改变 DataFrame 对象的索引，如下所示：

```
>>> data_df.index = ["a","b","c"]
>>> data_df
        country  density  median_age
a  United Kingdom     281          40
b          France     119          42
c           Italy     206          46
```

此时，如果我们想访问 country 列的第一个元素，可按如下操作进行：

```
>>> data_df["country"]["a"]
'United Kingdom'
```

同样地，如果想要改变 DataFrame 对象的列名，可以采取以下方法：

```
>>> data_df.columns = ["country_name","density","median_age"]
>>> data_df
    country_name  density  median_age
a  United Kingdom     281          40
b          France     119          42
c           Italy     206          46
```

假设已经将索引更改为字符串，则问题就是如何在索引未知的情况下访问第 n 行的元素。此时可利用 DataFrame 的 iloc() 方法，通过输入行号来访问第 n 行的元素，其用法示例如下：

```
>>> data_df.iloc[0]
country_name    United Kingdom
density                    281
median_age                  40
Name: a, dtype: object
>>> data_df.iloc[0:1]
    country_name  density  median_age
a  United Kingdom     281          40
>>> data_df.iloc[0:2]
    country_name  density  median_age
a  United Kingdom     281          40
b          France     119          42
>>> data_df.iloc[-1]
country_name    Italy
density           206
median_age         46
Name: c, dtype: object
```

可以看到，我们可以像访问列表一样访问 DataFrame 对象中的行。

在掌握了如何创建和访问 DataFrame 的基础上，下面将介绍如何向其中添加数据。假设需要在 DataFrame 中添加所有值均为 1 的列，可以按如下示例进行操作：

```
>>> data_df["ones"] = 1
>>> data_df
    country_name  density  median_age  ones
a  United Kingdom     281          40     1
b          France     119          42     1
c           Italy     206          46     1
```

我们可以通过以下几种方式删除列：

```
>>> del data_df["ones"]
>>> data_df
    country_name  density  median_age
```

```
a    United Kingdom    281    40
b            France    119    42
c             Italy    206    46
>>> data_df["ones"] = 1
>>> data_df
     country_name    density    median_age    ones
a    United Kingdom    281         40          1
b            France    119         42          1
c             Italy    206         46          1
>>> data_df.pop("ones")
a    1
b    1
c    1
Name: ones, dtype: int64
>>> data_df
     country_name    density    median_age
a    United Kingdom    281         40
b            France    119         42
c             Italy    206         46
```

在上面的程序中,首先使用了 del() 函数删除了名为 ones 的列。当再次添加后,又使用 pop() 方法删除了该列。值得注意的是,使用 del() 函数时是直接从 DataFrame 中进行删除,但使用 pop() 方法时,将返回被删除的列,就像从 DataFrame 中删除一样。

```
>>> data_df["ones"] = 1
>>> data_df["new_ones"] = data_df["ones"][1:2]
>>> data_df
     country_name    density    median_age    ones    new_ones
a    United Kingdom    281         40           1       NaN
b            France    119         42           1       1.0
c             Italy    206         46           1       NaN
>>> del data_df["ones"]
>>> del data_df["new_ones"]
```

可以看到,当使用部分列数据来构建一个新的列时,在 Pandas 中会用 NaN 值来填补空白。我们还可以给 DataFrame 对象插入新列并将其放在特定的位置,如下所示:

```
>>> data_df.insert(1, "twos", 2)
>>> data_df
     country_name    twos    density    median_age
a    United Kingdom    2       281         40
b            France    2       119         42
c             Italy    2       206         46
>>> del data_df["twos"]
```

在这里,我们创建一个新列,其值均为整数 2,命名为 twos,位置为 1(注意:位置 0 是第一个位置)。这样我们就能完全控制在 DataFrame 对象中何处进行列的添加操作。

到此为止，我们已经掌握了什么是 DataFrame，那么现在开始就可以进行一些更酷的操作。假设现在想获取值小于 200 的所有数据，则可以按照如下操作：

```
>>> data_df['density']< 200
a    False
b    True
c    False
Name: density, dtype: bool
>>> data_df[data_df['density']< 200]
   country_name  density  median_age
b        France      119          42
```

我们测试了 data_df 对象中 density 列中的值，看看哪些值小于 200。这个概念非常关键，该程序会测试该列中的每个元素，选出小于 200 的元素，并返回一个布尔列，显示哪些元素符合标准。随后将该列传递到 DataFrame 对象的方括号中，当条件为真时则返回值。我们可以对多个布尔语句执行此操作，此时所有真值对应的数据均可通过该语句返回，如下所示：

```
>>> data_df[data_df['density']< 250]
   country_name  density  median_age
b        France      119          42
c         Italy      206          46
>>> data_df["median_age"] > 42
a    False
b    False
c    True
Name: median_age, dtype: bool
>>> data_df[(data_df['density']< 250)
... &(data_df["median_age"]> 42)]
   country_name  density  median_age
c         Italy      206          46
```

需要注意的是，DataFrame 在此实例中并没有更改，它保持不变。如果要使用从此类操作返回的 DataFrame，则需将其分配给一个变量以供以后使用。

```
>>>(data_df['density']< 250) &(data_df["median_age"]> 42)
a    False
b    False
c    True
dtype: bool
>>> data_df['test'] = (data_df['density']< 250)
>>> data_df
   country_name   density  median_age   test
a  United Kingdom    281        40     False
b         France     119        42      True
c          Italy     206        46      True
>>> del data_df['test']
```

上述代码使用测试方法与上一个示例相同。将返回结果作为列 test 添加到 DataFrame 中。如果想利用 DataFrame 中的数据创建另一列，则可进行如下操作：

```
>>> data_df["density"].sum()
606
>>> data_df['density_proportion'] = data_df['density']/data_df['density'].sum()
>>> data_df
   country_name  density  median_age  density_proportion
a  United Kingdom   281       40         0.463696
b         France   119       42         0.196370
c          Italy   206       46         0.339934
```

上述代码中首先计算 density 列中的所有值的和，结果为 606，然后将 density 列的每一个值除以 606，最后获得新的一列。当然也可以对列数据执行标准的数学运算。下面就使用 NumPy 中的指数函数对列的各个元素进行指数运算：

```
>>> import numpy as np
>>> np.exp(data_df["density"])
a    1.088302e+122
b    4.797813e+51
c    2.915166e+89
Name: density, dtype: float64
```

也可以对 DataFrame 进行循环操作，就像之前在列表中的操作一样。

```
>>> for df in data_df:
...     df
...
'country_name'
'density'
'median_age'
'density_proportion'
```

只对列名之间进行循环并不是我们想要得到的结果，如果想真正对 DataFrame 进行深入地操作，则需引入转置的概念。

```
>>> data_df.T
                          a              b         c
country_name   United Kingdom      France     Italy
density                   281         119       206
median_age                 40          42        46
density_proportion   0.463696     0.19637  0.339934
```

通过转置将 DataFrame 转换为另一种形式，此时 DataFrame 的列变成了索引。为了循环访问 DataFrame 数据，可使用 iteritems() 方法，如下所示：

```
>>> for df in data_df.T.iteritems():
...     df
```

```
...
('a', country_name     United Kingdom
density                281
median_age             40
density_proportion     0.463696
Name: a, dtype: object)
('b', country_name     France
density                119
median_age             42
density_proportion     0.19637
Name: b, dtype: object)
('c', country_name     Italy
density                206
median_age             46
density_proportion     0.339934
Name: c, dtype: object)
```

可以看到,当在 DataFrame 上使用 iteritems() 方法时,每次循环均返回一个具有两个元素的元组,第一个元素是索引,第二个元素是存储在序列中行的值。为了更便捷地访问元组中的元素,可将元素分配给变量。示例如下:

```
>>> for ind, row in data_df.T.iteritems():
...     ind
...     row['country_name']
...
'a'
'United Kingdom'
'b'
'France'
'c'
'Italy'
```

上述示例中,我们将元组中的第一个元素赋值给变量 ind,将行序列赋值给变量 row,然后访问该行的 country_name 列并显示索引。此外,为了避免使用 DataFrame 的转置,也可以采用 iterrows() 方法直接访问 DataFrame 对象的行。

```
>>> for ind, row in data_df.iterrows():
...     ind
...     row
...
'a'
country_name           United Kingdom
density                281
median_age             40
density_proportion     0.463696
Name: a, dtype: object
'b'
country_name           France
density                119
median_age             42
```

```
density_proportion        0.19637
Name: b, dtype: object
'c'
country_name              Italy
density                   206
median_age                46
density_proportion        0.339934
Name: c, dtype: object
```

在介绍了如何向 DataFrame 中添加列后,下面继续讲解如何在 DataFrame 中添加行。可通过在 DataFrame 中使用 append()方法实现。具体的做法如下:

```
>>> data = [{"country": "United Kingdom", "density": 281, "median_age":40},
... {"country": "France", "density": 119, "median_age": 42},
... {"country":"Italy", "density":206, "median_age": 46}]
>>> data_df = pd.DataFrame(data)
>>> data_df
          country  density  median_age
0  United Kingdom      281          40
1          France      119          42
2           Italy      206          46
>>> new_row = [{"country": "Iceland", "median_age": 37, "density": 3}]
>>> new_row_data_df = pd.DataFrame(new_row)
>>> new_row_data_df
   country  median_age  density
0  Iceland          37        3
>>> data_df.append(new_row_data_df)
          country  density  median_age
0  United Kingdom      281          40
1          France      119          42
2           Italy      206          46
0         Iceland        3          37
>>> data_df
          country  density  median_age
0  United Kingdom      281          40
1          France      119          42
2           Italy      206          46
```

上述代码中,和先前一样,首先设置了初始数据,但这里使用了一个新的原始数据的副本。然后创建了 DataFrame 的一个对象 new_row,并将其传递到原始 DataFrame 的 append()方法[1]中,便可以得到 new_row 创建的 DataFrame,但其索引值为零,而这个索引在原始的 DataFrame 中本来就有。还可以看到,这些操作之后再次调用 DataFrame 时,没有看到使用 new_row 创建的数据。对于索引问题,我们可以利用 ignore_index 参

---

[1] DataFrame 中的 append()方法已经弃用,在 Pandas 未来版本中会移除,Pandas 建议使用 pandas.concat()方法,可以参考。

数来解决,具体操作如下所示:

```
>>> data_df.append(new_row_data_df, ignore_index = True)
         country    density    median_age
0  United Kingdom   281        40
1          France   119        42
2           Italy   206        46
3         Iceland   3          37
>>> data_df
         country    density    median_age
0  United Kingdom   281        40
1          France   119        42
2           Italy   206        46
```

从上面的程序可以看到,虽然索引已经做了排序,但为何 new_row 数据还是没有成为 DataFrame 的一部分呢?这主要是因为 append()方法不会改变原始的 DataFrame,因此,为了将 new_row 数据变成 DataFrame 的一部分,需要将 append()方法的结果赋值给一个新的 DataFrame 变量。可以为 DataFrame 重新分配一个具有相同名称的 data_df 变量,但这会导致原有数据的丢失,因此需要为 DataFrame 分配一个新的变量。

```
>>> new_data_df = data_df.append(new_row_data_df, ignore_index = True)
>>> new_data_df
         country    density    median_age
0  United Kingdom   281        40
1          France   119        42
2           Italy   206        46
3         Iceland   3          37
```

16.4 concat()、merge()和join()方法

concat()、merge()和join()是 DataFrame 的数据合并与连接方法。首先,让我们来看一下 DataFrame 中 concat()连接的概念。本节示例将使用 16.3 节中的数据创建的 DataFrame,并使用以下的 country 数据:

- 人口密度(density);
- 年龄中位数(median age);
- 人口(population,单位:百万);
- 人口变化(population change,单位:%)。

```
>>> import pandas as pd
>>> df1 = pd.DataFrame({"density": [119, 206, 240, 94],
...     "median_age": [42, 47, 46, 45],
...     "population": [65, 60, 83, 46],
```

```
...     "population_change": [0.22, -0.15, 0.32, 0.04]},
...     index=['France', 'Italy', 'Germany', 'Spain'])
>>> df2 = pd.DataFrame({"density": [153, 464, 36, 25],
...     "median_age": [38, 28, 38, 33],
...     "population": [1439, 1380, 331, 212],
...     "population_change": [0.39, 0.99, 0.59, 0.72]},
...     index=['China', 'India', 'USA', 'Brazil'])
>>> df3 = pd.DataFrame({"density": [9, 66, 347, 103],
...     "median_age": [40, 29, 48, 25],
...     "population": [145, 128, 126, 102],
...     "population_change": [0.04, 1.06, -0.30, 1.94]},
...     index=['Russia', 'Mexico', 'Japan', 'Egypt'])
>>> frames = [df1, df2, df3]
>>> result = pd.concat(frames)
>>> result
         density  median_age  population  population_change
France       119          42          65               0.22
Italy        206          47          60              -0.15
Germany      240          46          83               0.32
Spain         94          45          46               0.04
China        153          38        1439               0.39
India        464          28        1380               0.99
USA           36          38         331               0.59
Brazil        25          33         212               0.72
Russia         9          40         145               0.04
Mexico        66          29         128               1.06
Japan        347          48         126              -0.30
Egypt        103          25         102               1.94
```

上述程序中，我们创建了一个 DataFrame 的列表，并将其传递到 concat() 方法中，返回结果赋给 result 变量。result 变量是一个 DataFrame 对象，其列索引为 density、median_age、population 和 population_change，行索引为国家名。但是，如果没有示例中所示的索引值，将会怎样呢？

```
>>> df1 = pd.DataFrame({"density": [119, 206, 240, 94],
...     "median_age": [42, 47, 46, 45],
...     "population": [65, 60, 83, 46],
...     "population_change": [0.22, -0.15, 0.32, 0.04],
...     "country_name": ['France', 'Italy', 'Germany', 'Spain']})
>>> df2 = pd.DataFrame({"density": [153, 464, 36, 25],
...     "median_age": [38, 28, 38, 33],
...     "population": [1439, 1380, 331, 212],
...     "population_change": [0.39, 0.99, 0.59, 0.72],
...     "country_name": ['China', 'India', 'USA', 'Brazil']})
>>> df3 = pd.DataFrame({"density": [9, 66, 347, 103],
...     "median_age": [40, 29, 48, 25],
...     "population": [145, 128, 126, 102],
...     "population_change": [0.04, 1.06, -0.30, 1.94],
```

```
...     "country_name": ['Russia', 'Mexico', 'Japan', 'Egypt']})
>>> frames = [df1, df2, df3]
>>> result = pd.concat(frames)
>>> result
    density  median_age  population  population_change  country_name
0   119      42          65          0.22               France
1   206      47          60          -0.15              Italy
2   240      46          83          0.32               Germany
3   94       45          46          0.04               Spain
0   153      38          1439        0.39               China
1   464      28          1380        0.99               India
2   36       38          331         0.59               USA
3   25       33          212         0.72               Brazil
0   9        40          145         0.04               Russia
1   66       29          128         1.06               Mexico
2   347      48          126         -0.30              Japan
3   103      25          102         1.94               Egypt
```

可以看到，程序在创建 DataFrame 时为每个 DataFrame 均保留了值分别为 0、1、2、3 的索引。如果想将索引值更改为 0~11，则需要引入 ignore_index 参数，并将其值设置为 True。

```
>>> result = pd.concat(frames, ignore_index=True)
>>> result
    density  median_age  population  population_change  country_name
0   119      42          65          0.22               France
1   206      47          60          -0.15              Italy
2   240      46          83          0.32               Germany
3   94       45          46          0.04               Spain
4   153      38          1439        0.39               China
5   464      28          1380        0.99               India
6   36       38          331         0.59               USA
7   25       33          212         0.72               Brazil
8   9        40          145         0.04               Russia
9   66       29          128         1.06               Mexico
10  347      48          126         -0.30              Japan
11  103      25          102         1.94               Egypt
```

我们也可以对该示例进行扩展，即按之前的方法创建一个 DataFrame 列表，并将这些列表连接在一起。但需要使用参数 keys，并将其值设置为包含 region_one、region_two 和 region_three 的列表。

```
>>> df1 = pd.DataFrame({"density": [119, 206, 240, 94],
... "median_age": [42, 47, 46, 45],
... "population": [65, 60, 83, 46],
... "population_change": [0.22, -0.15, 0.32, 0.04],
... "country_name": ['France', 'Italy', 'Germany', 'Spain']})
```

```
>>> df2 = pd.DataFrame({"density": [153, 464, 36, 25],
... "median_age": [38, 28, 38, 33],
... "population": [1439, 1380, 331, 212],
... "population_change": [0.39, 0.99, 0.59, 0.72],
... "country_name": ['China', 'India', 'USA', 'Brazil']})
>>> df3 = pd.DataFrame({"density": [9, 66, 347, 103],
... "median_age": [40, 29, 48, 25],
... "population": [145, 128, 126, 102],
... "population_change": [0.04, 1.06, -0.30, 1.94],
... "country_name": ['Russia', 'Mexico', 'Japan', 'Egypt']})
>>> frames = [df1, df2, df3]
>>> result = pd.concat(frames, keys = ["region_one", "region_two", "region_three"])
>>> result
              density  median_age  population  population_change  country_name
region_one   0  119       42          65         0.22              France
             1  206       47          60        -0.15              Italy
             2  240       46          83         0.32              Germany
             3   94       45          46         0.04              Spain
region_two   0  153       38        1439         0.39              China
             1  464       28        1380         0.99              India
             2   36       38         331         0.59              USA
             3   25       33         212         0.72              Brazil
region_three 0    9       40         145         0.04              Russia
             1   66       29         128         1.06              Mexico
             2  347       48         126        -0.30              Japan
             3  103       25         102         1.94              Egypt
>>> result.loc['region_two']
   density  median_age  population  population_change  country_name
0  153      38          1439        0.39               China
1  464      28          1380        0.99               India
2   36      38           331        0.59               USA
3   25      33           212        0.72               Brazil
```

运行代码后可以看到,通过传入 keys 参数,我们可以拥有 DataFrame 的其他层级,没有再显示前面示例中的索引,在 concat() 连接中允许选择其中一个 DataFrame。当查看 result 变量的索引时,会得到如下结果[1]:

```
>>> result.index
MultiIndex(levels = [['region_one', 'region_two', 'region_three'],
[0, 1, 2, 3]],
codes = [[0, 0, 0, 0, 1, 1, 1, 1, 2, 2, 2, 2],
[0, 1, 2, 3, 0, 1, 2, 3, 0, 1, 2, 3]])
#译者运行结果
>>> result.index
```

---

[1] 可能由于版本原因,译者运行结果与原书中的结果不一致,为了与文中数字相对应,文中采用原文结果,但将译者运行结果也放入代码中,读者可以自行验证。

```
MultiIndex([( 'region_one', 0),
            ( 'region_one', 1),
            ( 'region_one', 2),
            ( 'region_one', 3),
            ( 'region_two', 0),
            ( 'region_two', 1),
            ( 'region_two', 2),
            ( 'region_two', 3),
            ('region_three', 0),
            ('region_three', 1),
            ('region_three', 2),
            ('region_three', 3)],
           )
```

这种索引常被称为多级索引(multilevel index),其作用是给出每个元素的索引值。本例中,索引级别分别为['region_one','region_two','region_three']和[0,1,2,3]。可使用标签 codes 确定每行的索引,该标签有两个列表,第一个列表中的 0、1、2 分别对应region_one、region_two、region_three,而第二个列表中的 0、1、2、3 分别对应级别 0、1、2、3。我们可以用 names 参数命名这些级别,示例如下:

```
>>> result = pd.concat(frames, keys = ["region_one", "region_two",
... "region_three"], names = ["region","item"])
>>> result
                    density  median_age  population  population_change  country_name
region       item
region_one   0      119      42          65          0.22               France
             1      206      47          60          -0.15              Italy
             2      240      46          83          0.32               Germany
             3      94       45          46          0.04               Spain
region_two   0      153      38          1439        0.39               China
             1      464      28          1380        0.99               India
             2      36       38          331         0.59               USA
             3      25       33          212         0.72               Brazil
region_three 0      9        40          145         0.04               Russia
             1      66       29          128         1.06               Mexico
             2      347      48          126         -0.30              Japan
             3      103      25          102         1.94               Egypt
```

在前面示例中,我们使用 concat() 方法将 DataFrame 数据连接在一起。下面将给出concat()的另一种应用方法,通过将 France、Italy、Argentina 及 Thailand 的城市人口比例连接到原始的 DataFrame 中来演示。

```
>>> df1 = pd.DataFrame({"density": [119, 206, 240, 94],
...     "median_age": [42, 47, 46, 45],
...     "population": [65, 60, 83, 46],
...     "population_change": [0.22, -0.15, 0.32, 0.04]},
```

```
...        index = ['France', 'Italy', 'Germany', 'Spain'])
>>> df4 = pd.DataFrame({"urban_population": [82, 69, 93, 51]},
...        index = ['France', 'Italy', 'Argentina', 'Thailand'])
>>> result = pd.concat([df1, df4], axis = 1, sort = False)
>>> result
          density  median_age  population  population_change  urban_population
France    119.0    42.0        65.0        0.22               82.0
Italy     206.0    47.0        60.0        -0.15              69.0
Germany   240.0    46.0        83.0        0.32               NaN
Spain     94.0     45.0        46.0        0.04               NaN
Argentina NaN      NaN         NaN         NaN                93.0
Thailand  NaN      NaN         NaN         NaN                51.0
```

在该程序中,与前面示例一样,对 DataFrame 的列表使用了 concat() 方法连接,但现在传入了 axis=1 的参数。axis 参数表示连接操作的轴,axis=0 表示行,axis=1 表示列。因此,当寻找到 France 和 Italy 索引中的共性后,即可添加一新的列,并给该列填充相应的值,没有匹配值的元素填入 NaN。此外,设置 sort = False,表示将两个值按顺序进行连接。如果设 sort = True,则结果如下:

```
>>> result = pd.concat([df1, df4], axis = 1, sort = True)
>>> result
          density  median_age  population  population_change  urban_population
Argentina NaN      NaN         NaN         NaN                93.0
France    119.0    42.0        65.0        0.22               82.0
Germany   240.0    46.0        83.0        0.32               NaN
Italy     206.0    47.0        60.0        -0.15              69.0
Spain     94.0     45.0        46.0        0.04               NaN
Thailand  NaN      NaN         NaN         NaN                51.0
```

可以看到,当设 sort = True 时,结果所得的值则是按行索引进行了排序。当设 axis = 0 时,以上的程序的运行结果又如何呢?

```
>>> result = pd.concat([df1, df4], axis = 0, sort = False)
>>> result
          density  median_age  population  population_change  urban_population
France    119.0    42.0        65.0        0.22               NaN
Italy     206.0    47.0        60.0        -0.15              NaN
Germany   240.0    46.0        83.0        0.32               NaN
Spain     94.0     45.0        46.0        0.04               NaN
France    NaN      NaN         NaN         NaN                82.0
Italy     NaN      NaN         NaN         NaN                69.0
Argentina NaN      NaN         NaN         NaN                93.0
Thailand  NaN      NaN         NaN         NaN                51.0
```

上述所做的只是将 DataFrame 数据按顺序将一个放在另一个的下面连接起来,其中 France 和 Italy 的索引是重复的。

concat() 方法中还有一个附加参数 join,下面研究 join 参数及其取值的用法。

```
>>> df1 = pd.DataFrame({"density": [119, 206, 240, 94],
... "median_age": [42, 47, 46, 45],
... "population": [65, 60, 83, 46],
... "population_change": [0.22, -0.15, 0.32, 0.04]},
... index = ['France', 'Italy', 'Germany', 'Spain'])
>>> df4 = pd.DataFrame({"urban_population": [82, 69, 93, 51]},
... index = ['France', 'Italy', 'Argentina', 'Thailand'])
>>> result = pd.concat([df1, df4], axis = 1, join = "inner")
>>> result
        density  median_age  population  population_change  urban_population
France      119          42          65               0.22                82
Italy       206          47          60              -0.15                69
```

程序运行结果只返回了两行，如果回顾之前的示例时可以看到，只有这两行在其 DataFrame 所有列上都有数值。join="inner"与数据库的 join 操作类似，后续内容会有介绍，但这里没有指定其应用的键值。

接下来，添加参数 join_axes[①]，并将其设为 df1.index。

```
>>> df1 = pd.DataFrame({"density": [119, 206, 240, 94],
... "median_age": [42, 47, 46, 45],
... "population": [65, 60, 83, 46],
... "population_change": [0.22, -0.15, 0.32, 0.04]},
... index = ['France', 'Italy', 'Germany', 'Spain'])
>>> df4 = pd.DataFrame({"urban_population": [82, 69, 93, 51]},
... index = ['France', 'Italy', 'Argentina', 'Thailand'])
>>> result = pd.concat([df1, df4], axis = 1, join_axes = [df1.index])
>>> result
         density  median_age  population  population_change  urban_population
France       119          42          65               0.22              82.0
Italy        206          47          60              -0.15              69.0
Germany      240          46          83               0.32               NaN
Spain         94          45          46               0.04               NaN
```

可以看到，这里只返回了 df1 中索引中的值，并显示了参数 axis=1 中的所有列。默认情况下，join_axes 设置为 False。

接下来，可使用以下参数来忽略索引：

```
>>> df1 = pd.DataFrame({"density": [119, 206, 240, 94],
... "median_age": [42, 47, 46, 45],
... "population": [65, 60, 83, 46],
... "population_change": [0.22, -0.15, 0.32, 0.04]},
... index = ['France', 'Italy', 'Germany', 'Spain'])
>>> df4 = pd.DataFrame({"urban_population": [82, 69, 93, 51]},
```

---

① 新版本 Pandas 已经删除了 join_axes，按照这段程序调试可能会报错，可以改用 merge() 方法将两个 DataFrame 按照同一列合并。

```
... index = ['France', 'Italy', 'Argentina', 'Thailand'])
>>> result = pd.concat([df1, df4], ignore_index = True, sort = True)
>>> result
   density  median_age  population  population_change  urban_population
0    119.0        42.0        65.0               0.22               NaN
1    206.0        47.0        60.0              -0.15               NaN
2    240.0        46.0        83.0               0.32               NaN
3     94.0        45.0        46.0               0.04               NaN
4      NaN         NaN         NaN                NaN              82.0
5      NaN         NaN         NaN                NaN              69.0
6      NaN         NaN         NaN                NaN              93.0
7      NaN         NaN         NaN                NaN              51.0
```

上述程序运行后可看到，结果中丢失了 df1 和 df2 的索引值，非数的值均填入"NaN"。

我们也可以直接在 DataFrame 中使用 append() 方法来获得相同的结果。

```
>>> df1 = pd.DataFrame({"density": [119, 206, 240, 94],
... "median_age": [42, 47, 46, 45],
... "population": [65, 60, 83, 46],
... "population_change": [0.22, -0.15, 0.32, 0.04]},
... index = ['France', 'Italy', 'Germany', 'Spain'])
>>> df4 = pd.DataFrame({"urban_population": [82, 69, 93, 51]},
... index = ['France', 'Italy', 'Argentina', 'Thailand'])
>>> result = df1.append(df4, ignore_index = True, sort = True)
>>> result
   density  median_age  population  population_change  urban_population
0    119.0        42.0        65.0               0.22               NaN
1    206.0        47.0        60.0              -0.15               NaN
2    240.0        46.0        83.0               0.32               NaN
3     94.0        45.0        46.0               0.04               NaN
4      NaN         NaN         NaN                NaN              82.0
5      NaN         NaN         NaN                NaN              69.0
6      NaN         NaN         NaN                NaN              93.0
7      NaN         NaN         NaN                NaN              51.0
```

concat() 操作不仅适用于 DataFrame，也适用于 Series(序列)。

```
>>> df1 = pd.DataFrame({"density": [119, 206, 240, 94],
... "median_age": [42, 47, 46, 45],
... "population": [65, 60, 83, 46],
... "population_change": [0.22, -0.15, 0.32, 0.04]},
... index = ['France', 'Italy', 'Germany', 'Spain'])
>>> s1 = pd.Series([82, 69, 93, 51],
... index = ['France', 'Italy', 'Germany', 'Spain'],
... name = "urban_population")
>>> s1
```

```
France      82
Italy       69
Germany     93
Spain       51
Name: urban_population, dtype: int64
>>> result = pd.concat([df1, s1], axis = 1)
>>> result
         density  median_age  population  population_change  urban_population
France       119          42          65               0.22                82
Italy        206          47          60              -0.15                69
Germany      240          46          83               0.32                93
Spain         94          45          46               0.04                51
```

值得注意的是,在序列 s1 中添加了一个 name 项,当 DataFrame 与 Series 合并时,name 值则成为一个列的名称。我们还可以在 concat()列表中传递多个序列,下面演示添加第二个序列,内容为世界份额百分比(world share percentage)。

```
>>> df1 = pd.DataFrame({"density": [119, 206, 240, 94],
... "median_age": [42, 47, 46, 45],
... "population": [65, 60, 83, 46],
... "population_change": [0.22, -0.15, 0.32, 0.04]},
... index = ['France', 'Italy', 'Germany', 'Spain'])
>>> s1 = pd.Series([82, 69, 93, 51],
... index = ['France', 'Italy', 'Germany', 'Spain'],
... name = "urban_population")
>>> s2 = pd.Series([0.84, 0.78, 1.07, 0.60],
... index = ['France', 'Italy', 'Germany', 'Spain'],
... name = "world_share")
>>> s2
France     0.84
Italy      0.78
Germany    1.07
Spain      0.60
Name: world_share, dtype: float64
>>> result = pd.concat([df1, s1, s2], axis = 1)
>>> result
         density  median_age  ......  urban_population  world_share
France       119          42  ......                82         0.84
Italy        206          47  ......                69         0.78
Germany      240          46  ......                93         1.07
Spain         94          45  ......                51         0.60
[4 rows x 6 columns]
```

接下来,将序列以列表的形式传递给 concat()创建 DataFrame,并通过指定 keys 参数对列进行重新命名。

```
>>> s1 = pd.Series([82, 69, 93, 51],
... index = ['France', 'Italy', 'Germany', 'Spain'],
```

```
...   name = "urban_population")
>>> s2 = pd.Series([0.84, 0.78, 1.07, 0.60],
...   index = ['France', 'Italy', 'Germany', 'Spain'],
...   name = "world_share")
>>> pd.concat([s1, s2], axis = 1, keys = ["urban_population","world_share"])
         urban_population    world_share
France        82               0.84
Italy         69               0.78
Germany       93               1.07
Spain         51               0.60
```

接下来,将前面创建的 3 个 DataFrame 赋值到一个字典中,并为每个 DataFrame 设置一个键,最后将字典传递给 concat(),如下所示:

```
>>> df1 = pd.DataFrame({"density": [119, 206, 240, 94],
...   "median_age": [42, 47, 46, 45],
...   "population": [65, 60, 83, 46],
...   "population_change": [0.22, -0.15, 0.32, 0.04]},
...   index = ['France', 'Italy', 'Germany', 'Spain'])
>>> df2 = pd.DataFrame({"density": [153, 464, 36, 25],
...   "median_age": [38, 28, 38, 33],
...   "population": [1439, 1380, 331, 212],
...   "population_change": [0.39, 0.99, 0.59, 0.72]},
...   index = ['China', 'India', 'USA', 'Brazil'])
>>> df3 = pd.DataFrame({"density": [9, 66, 347, 103],
...   "median_age": [40, 29, 48, 25],
...   "population": [145, 128, 126, 102],
...   "population_change": [0.04, 1.06, -0.30, 1.94]},
...   index = ['Russia', 'Mexico', 'Japan', 'Egypt'])
>>> pieces = {"region_one": df1, "region_two": df2, "region_three": df3}
>>> result = pd.concat(pieces)
>>> result
                      density   median_age   population   population_change
region_one   France      119        42           65           0.22
             Italy       206        47           60          -0.15
             Germany     240        46           83           0.32
             Spain        94        45           46           0.04
region_two   China       153        38         1439           0.39
             India       464        28         1380           0.99
             USA          36        38          331           0.59
             Brazil       25        33          212           0.72
region_three Russia        9        40          145           0.04
             Mexico       66        29          128           1.06
             Japan       347        48          126          -0.30
             Egypt       103        25          102           1.94
```

在使用字典时,自动创建了一个具有多级索引的 DataFrame,其中第一级是字典的键,第二级是 DataFrame 的索引。下面的方法与上例类似,不同的是通过 keys 参数传入了一个可选的键列表。

```
>>> df1 = pd.DataFrame({"density": [119, 206, 240, 94],
... "median_age": [42, 47, 46, 45],
... "population": [65, 60, 83, 46],
... "population_change": [0.22, -0.15, 0.32, 0.04]},
... index = ['France', 'Italy', 'Germany', 'Spain'])
>>> df2 = pd.DataFrame({"density": [153, 464, 36, 25],
... "median_age": [38, 28, 38, 33],
... "population": [1439, 1380, 331, 212],
... "population_change": [0.39, 0.99, 0.59, 0.72]},
... index = ['China', 'India', 'USA', 'Brazil'])
>>> df3 = pd.DataFrame({"density": [9, 66, 347, 103],
... "median_age": [40, 29, 48, 25],
... "population": [145, 128, 126, 102],
... "population_change": [0.04, 1.06, -0.30, 1.94]},
... index = ['Russia', 'Mexico', 'Japan', 'Egypt'])
>>> pieces = {"region_one": df1, "region_two": df2, "region_three": df3}
>>> result = pd.concat(pieces, keys = ["region_two", "region_three"])
>>> result
                       density   median_age   population   population_change
region_two    China    153       38           1439         0.39
              India    464       28           1380         0.99
              USA      36        38           331          0.59
              Brazil   25        33           212          0.72
region_three  Russia   9         40           145          0.04
              Mexico   66        29           128          1.06
              Japan    347       48           126          -0.30
              Egypt    103       25           102          1.94
```

在研究了concat()和append()方法之后，现在考虑pandas如何处理数据库样式合并，可通过merge()方法实现。我们将通过示例解释每个连接类型的细节，但有必要先来了解一下数据库连接的基础知识。当谈到数据库样式连接(database style join)时，指的是通过公共值将表连接在一起的机制。表的形式将取决于连接的类型，后续将以示例的形式给出数据库的左(left)、右(right)、外部(outer)和内部(inner)连接的示例。

我们将以国家数据为例实现DataFrame数据的合并，在包含国家通用数据的DataFrame的基础上，添加与国家相关的世界份额比例数据。

```
>>> left = pd.DataFrame({"density": [119, 206, 240, 94],
...                      "median_age": [42, 47, 46, 45],
...                      "population": [65, 60, 83, 46],
...                      "population_change": [0.22, -0.15, 0.32, 0.04],
...                      "country": ['France', 'Italy', 'Germany', 'Spain']})
>>> right = pd.DataFrame({"world_share": [0.84, 0.78, 1.07, 0.60],
...                       "country": ['France', 'Italy', 'Germany', 'Spain']})
>>> result = pd.merge(left, right, on = "country")
>>> result
```

```
     density  median_age  population  population_change  country  world_share
0    119      42          65          0.22               France   0.84
1    206      47          60          -0.15              Italy    0.78
2    240      46          83          0.32               Germany  1.07
3    94       45          46          0.04               Spain    0.60
```

在本例中,以国家名称作为公共键将两个 DataFrame 连接起来,得到的结果是一个只有一个国家列的 DataFrame,左边和右边的列都是通过 merge() 方法合并进来的。

下面将研究对人进行左、右连接对人的 merge() 方法,与上例不同的是这次将有两个公共键,它们将作为列表传递给 on 参数。这允许我们在多个值相同的情况下进行连接操作。

```
>>> left = pd.DataFrame({"density": [119, 206, 240, 94],
...         "median_age": [42, 47, 46, 45],
...         "population": [65, 60, 83, 46],
...         "population_change": [0.22, -0.15, 0.32, 0.04],
...         "country": ['France', 'Italy', 'Germany', 'Spain']})
>>> right = pd.DataFrame({"world_share": [0.84, 0.78, 1.07, 0.60],
...         "population": [65, 60, 85, 46],
...         "country": ['France', 'Italy', 'Germany', 'Spain']})
>>> result = pd.merge(left, right, on=["country", "population"])
>>> result
     density  median_age  population  population_change  country  world_share
0    119      42          65          0.22               France   0.84
1    206      47          60          -0.15              Italy    0.78
2    94       45          46          0.04               Spain    0.60
```

上述代码基于公共键 country 和 population 实现了连接,得到的 DataFrame 是两个 DataFrame 共享相同的 country 和 population。这里丢失了每个 DataFrame 中的一行数据,因为不能在两者上共享 country 和 population。接下来运行相同的代码,并添加一个 how 参数,该参数等于 left。

```
>>> left = pd.DataFrame({"density": [119, 206, 240, 94],
...         "median_age": [42, 47, 46, 45],
...         "population": [65, 60, 83, 46],
...         "population_change": [0.22, -0.15, 0.32, 0.04],
...         "country": ['France', 'Italy', 'Germany', 'Spain']})
>>> right = pd.DataFrame({"world_share": [0.84, 0.78, 1.07, 0.60],
...         "country": ['France', 'Italy', 'Germany', 'Spain']})
>>> result = pd.merge(left, right, how="left", on="country")
>>> result
     density  median_age  population  population_change  country  world_share
0    119      42          65          0.22               France   0.84
1    206      47          60          -0.15              Italy    0.78
2    240      46          83          0.32               Germany  1.07
3    94       45          46          0.04               Spain    0.60
```

其结果就是所谓的左连接(left join)。因此,我们保留左侧 DataFrame 的所有信息,仅保留右侧 DataFrame 中具有与左侧相同键的元素。在这种情况下,我们保留左侧 DataFrame 中的所有信息。

接下来,我们使用与前面相同的示例来看一下右连接(right join)。

```
>>> left = pd.DataFrame({"density": [119, 206, 240, 94],
...         "median_age": [42, 47, 46, 45],
...         "population": [65, 60, 83, 46],
...         "population_change": [0.22, -0.15, 0.32, 0.04],
...         "country": ['France', 'Italy', 'Germany', 'Spain']})
>>> right = pd.DataFrame({"world_share": [0.84, 0.78, 1.07, 0.60],
...         "population": [65, 60, 85, 46],
...         "country": ['France', 'Italy', 'Germany', 'Spain']})
>>> result = pd.merge(left, right, how = "right", on = ["country", "population"])
>>> result
    density  median_age  population  population_change  country  world_share
0   119.0    42.0        65          0.22               France   0.84
1   206.0    47.0        60          -0.15              Italy    0.78
2   NaN      NaN         85          NaN                Germany  1.07
3   94.0     45.0        46          0.04               Spain    0.60
```

本质上,右连接与左连接相同,但现在是左侧 DataFrame 连接到右侧的 DataFrame 上,这与之前看到的左连接正好相反。

下一个要考虑的连接是外部连接(outer join),为了完整起见,我们再次使用前面的示例来说明它是如何工作的。

```
>>> left = pd.DataFrame({"density": [119, 206, 240, 94],
...         "median_age": [42, 47, 46, 45],
...         "population": [65, 60, 83, 46],
...         "population_change": [0.22, -0.15, 0.32, 0.04],
...         "country": ['France', 'Italy', 'Germany', 'Spain']})
>>> right = pd.DataFrame({"world_share": [0.84, 0.78, 1.07, 0.60],
...         "population": [65, 60, 85, 46],
...         "country": ['France', 'Italy', 'Germany', 'Spain']})
>>> result = pd.merge(left, right, how = "outer", on = ["country", "population"])
>>> result
    density  median_age  population  population_change  country  world_share
0   119.0    42.0        65          0.22               France   0.84
1   206.0    47.0        60          -0.15              Italy    0.78
2   240.0    46.0        83          0.32               Germany  NaN
3   94.0     45.0        46          0.04               Spain    0.60
4   NaN      NaN         85          NaN                Germany  1.07
```

外部连接是左连接和右连接的组合,因此得到的行数比每个 DataFrame 中的行数多,因为左连接和右连接给出的结果不同,所有这些都包含在外部连接结果中。

我们考虑的最后一个连接是内部连接(inner join)。

```
>>> left = pd.DataFrame({"density": [119, 206, 240, 94],
...                     "median_age": [42, 47, 46, 45],
...                     "population": [65, 60, 83, 46],
...                     "population_change": [0.22, -0.15, 0.32, 0.04],
...                     "country": ['France', 'Italy', 'Germany', 'Spain']})
>>> right = pd.DataFrame({"world_share": [0.84, 0.78, 1.07, 0.60],
...                      "population": [65, 60, 85, 46],
...                      "country": ['France', 'Italy', 'Germany', 'Spain']})
>>> result = pd.merge(left, right, how="inner", on=["country", "population"])
>>> result
   density  median_age  population  population_change  country  world_share
0      119          42          65               0.22   France         0.84
1      206          47          60              -0.15    Italy         0.78
2       94          45          46               0.04    Spain         0.60
```

内部连接只给出了在左、右 DataFrame 上都具有的通用性的结果，这也是 how 参数的默认值，如下所示：

```
>>> result = pd.merge(left, right, on=["country", "population"])
>>> result
   density  median_age  population  population_change  country  world_share
0      119          42          65               0.22   France         0.84
1      206          47          60              -0.15    Italy         0.78
2       94          45          46               0.04    Spain         0.60
```

接下来，使用 population 和 country 列连接两个 DataFrame，但仅在 country 列上使用外部连接进行连接：

```
>>>
>>> left = pd.DataFrame({"population": [65, 60, 83, 46],
...                     "country": ['France', 'Italy', 'Germany', 'Spain']})
>>> right = pd.DataFrame({"population": [65, 60, 85, 46],
...                      "country": ['France', 'Italy', 'Germany', 'Spain']})
>>> result = pd.merge(left, right, on="country", how="outer")
>>> result
   population_x  country  population_y
0            65   France            65
1            60    Italy            60
2            83  Germany            85
3            46    Spain            46
```

由上可见，如果列相同且未在连接中使用，则名称会更改。现在有 population_x 和 population_y，如果要对 population 列进行操作，可能会出现问题。因此，需要对两者进行区分，而 Pandas 也要对此进行处理。

接下来，通过设置 indicator 选项为 True 进行连接操作。这里，有两个 DataFrame，只有一个 country 列要合并，下面希望使用外部连接来执行此项操作：

```
>>> left = pd.DataFrame({"population": [65, 60, 83, 46],
...         "country": ['France', 'Italy', 'Germany', 'Spain']})
>>> right = pd.DataFrame({"population": [65, 60, 85, 46],
...         "country": ['France', 'Italy', 'Germany', 'Spain']})
>>> result = pd.merge(left, right, on = "country", how = "outer", indicator = True)
>>> result
   population_x   country  population_y  _merge
0         65      France        65       both
1         60      Italy         60       both
2         83      Germany       85       both
3         46      Spain         46       both
```

上述程序结果显示的连接是根据索引位置逐个完成的，因此可以是左连接、右连接或两者都是。在这里，我们看到从一个到另一个的连接都是在两者上完成的。

merge()方法是实现两个DataFrame合并的Pandas方法，但是我们也可以使用DataFrame的join()方法将一个连接到另一个。

```
>>> left = pd.DataFrame({"density": [ 119, 206, 240, 94 ],
...         "median_age": [ 42, 47, 46, 45 ],
...         "population": [ 65, 60, 83, 46 ],
...         "population_change": [ 0.22, -0.15, 0.32, 0.04 ]},
...         index = [ 'France', 'Italy', 'Germany', 'Spain' ])
>>> right = pd.DataFrame({"world_share": [ 0.84, 0.78, 1.07, 0.60 ]},
...         index = [ 'France', 'Italy', 'Germany',
...         'United Kingdom' ])
>>> result = left.join(right)
>>> result
         density  median_age  population  population_change  world_share
France    119       42         65          0.22              0.84
Italy     206       47         60         -0.15              0.78
Germany   240       46         83          0.32              1.07
Spain      94       45         46          0.04              NaN
```

我们看到的是，左侧的DataFrame被保留，连接右侧的DataFrame，其中右侧的键与左侧的键相匹配。像merge()一样，join()方法可以选择如何连接DataFrame，可以像前面那样指定连接的选项。使用前面的相同示例，展示如下：

```
>>> left = pd.DataFrame({"density": [119, 206, 240, 94],
...         "median_age": [42, 47, 46, 45],
...         "population": [65, 60, 83, 46],
...         "population_change": [0.22, -0.15, 0.32, 0.04]},
...         index = ['France', 'Italy', 'Germany', 'Spain'])
>>> right = pd.DataFrame({"world_share": [0.84, 0.78, 1.07, 0.60]},
...         index = ['France', 'Italy', 'Germany',
...         'United Kingdom'])
>>> result = left.join(right, how = "outer")
>>> result
```

```
              density  median_age  population  population_change  world_share
France          119.0        42.0        65.0               0.22         0.84
Germany         240.0        46.0        83.0               0.32         1.07
Italy           206.0        47.0        60.0              -0.15         0.78
Spain            94.0        45.0        46.0               0.04          NaN
United Kingdom    NaN         NaN         NaN                NaN         0.60
```

在使用外部连接时,保留了两个 DataFrame 的所有信息,正如在使用 merge() 方法时的做法一样,如果其中一个 DataFrame 中没有值,则使用 NaN 进行填充。同样地,对于使用内部连接的相同示例结果如下:

```
>>> left = pd.DataFrame({"density": [119, 206, 240, 94],
...          "median_age": [42, 47, 46, 45],
...          "population": [65, 60, 83, 46],
...          "population_change": [0.22, -0.15, 0.32, 0.04]},
...          index = ['France', 'Italy', 'Germany', 'Spain'])
>>> right = pd.DataFrame({"world_share": [0.84, 0.78, 1.07, 0.60]},
...          index = ['France', 'Italy', 'Germany',
...          'United Kingdom'])
>>> result = left.join(right, how = "inner")
>>> result
         density  median_age  population  population_change  world_share
France       119          42          65               0.22         0.84
Italy        206          47          60              -0.15         0.78
Germany      240          46          83               0.32         1.07
>>>
```

正如预期的那样,内部连接只保留了 DataFrame 的共同数据,这里是只包含 France、Italy 和 Germany 的索引。

如果将一些不同的参数传递给 merge() 方法,就可以在不使用 how 参数的情况下获得相同的结果。此时,需要将参数 left_index 和 right_index 的值设置为 True。我们得到了与 join() 方法相同的行为,其设置方式为 inner。

```
>>> left = pd.DataFrame({"density": [119, 206, 240, 94],
...          "median_age": [42, 47, 46, 45],
...          "population": [65, 60, 83, 46],
...          "population_change": [0.22, -0.15, 0.32, 0.04]},
...          index = ['France', 'Italy', 'Germany', 'Spain'])
>>> right = pd.DataFrame({"world_share": [0.84, 0.78, 1.07, 0.60]},
...          index = ['France', 'Italy', 'Germany',
...          'United Kingdom'])
>>> result = pd.merge(left, right, left_index = True, right_index = True,)
>>> result
         density  median_age  population  population_change  world_share
France       119          42          65               0.22         0.84
Italy        206          47          60              -0.15         0.78
Germany      240          46          83               0.32         1.07
```

接下来，对左侧 DataFrame 应用 join() 方法，并使用 on 参数进行连接，示例如下：

```
>>> left = pd.DataFrame({"density": [119, 206, 240, 94],
...     "median_age": [42, 47, 46, 45],
...     "population": [65, 60, 83, 46],
...     "population_change": [0.22, -0.15, 0.32, 0.04],
...     "country": ['France', 'Italy', 'Germany', 'Spain']})
>>> right = pd.DataFrame({"world_share": [0.84, 0.78, 1.07, 0.60]},
...     index = ['France', 'Italy', 'Germany',
...     'United Kingdom'])
>>> left
   density  median_age  population  population_change  country
0  119      42          65          0.22               France
1  206      47          60          -0.15              Italy
2  240      46          83          0.32               Germany
3  94       45          46          0.04               Spain
>>> right
                world_share
France          0.84
Italy           0.78
Germany         1.07
United Kingdom  0.60
>>> result = left.join(right, on = "country")
>>> result
   density  median_age  population  population_change  country  world_share
0  119      42          65          0.22               France   0.84
1  206      47          60          -0.15              Italy    0.78
2  240      46          83          0.32               Germany  1.07
3  94       45          46          0.04               Spain    NaN
```

在 join() 方法中，on 参数指定的 country 列是右侧 DataFrame 连接的列索引，输出结果为一个由第一个 DataFrame 索引的 DataFrame。当使用数据库时，可能会看到这种类型的方法，并且希望在另一个表中相应值的列 id 上进行连接。这个示例可以扩展到 on 参数有多个值的情况，但需要多级索引才能做到这一点，这将在本书后面介绍。我们可以通过添加 how 参数并将其设置为 inner 来移除任何 NaN 值，执行的内部连接如下所示：

```
>>> left = pd.DataFrame({"density": [119, 206, 240, 94],
...     "median_age": [42, 47, 46, 45],
...     "population": [65, 60, 83, 46],
...     "population_change": [0.22, -0.15, 0.32, 0.04],
...     "country": ['France', 'Italy', 'Germany', 'Spain']})
>>> right = pd.DataFrame({"world_share": [0.84, 0.78, 1.07, 0.60]},
...     index = ['France', 'Italy', 'Germany',
...     'United Kingdom'])
>>> result = left.join(right, on = "country", how = "inner")
>>> result
```

	density	median_age	population	population_change	country	world_share
0	119	42	65	0.22	France	0.84
1	206	47	60	-0.15	Italy	0.78
2	240	46	83	0.32	Germany	1.07

接下来要考虑的是丢失数据这一重要概念。我们都希望使用完美的数据集，但现实通常不会这样，拥有处理丢失或错误数据的能力十分重要。幸运的是，Pandas 为解决该问题提供一些非常有用的工具。首先展示如何识别数据集中的 NaN，示例代码如下：

```
>>> left = pd.DataFrame({"density": [119, 206, 240, 94],
...         "median_age": [42, 47, 46, 45],
...         "population": [65, 60, 83, 46],
...         "population_change": [0.22, -0.15, 0.32, 0.04],
...         "country": ['France', 'Italy', 'Germany', 'Spain']})
>>> right = pd.DataFrame({"world_share": [0.84, 0.78, 1.07, 0.60],
...         "population": [65, 60, 85, 46],
...         "country": ['France', 'Italy', 'Germany', 'Spain']})
>>> result = pd.merge(left, right, how="outer", on=["country", "population"])
>>> result
   density  median_age  population  population_change  country  world_share
0   119.0       42.0          65           0.22         France      0.84
1   206.0       47.0          60          -0.15         Italy       0.78
2   240.0       46.0          83           0.32         Germany     NaN
3    94.0       45.0          46           0.04         Spain       0.60
4    NaN        NaN           85           NaN          Germany     1.07
>>> pd.isna(result['density'])
0    False
1    False
2    False
3    False
4    True
Name: density, dtype: bool
>>> result['median_age'].notna()
0    True
1    True
2    True
3    True
4    False
Name: median_age, dtype: bool
>>> result.isna()
   density  median_age  population  population_change  country  world_share
0   False    False       False       False              False    False
1   False    False       False       False              False    False
2   False    False       False       False              False    True
3   False    False       False       False              False    False
4   True     True        False       True               False    False
```

上面示例中，我们使用了前面用到的 DataFrame。使用 merge() 方法，并将其中的

how 参数设置为 outer,从而合并得到了一个 DataFrame(其中具有 NaN 值)。随后展示如何找到这些值在 DataFrame 中的位置。首先采用 Pandas 库的 isna()方法,它可以测试 DataFrame 列上的每个元素是否为 NaN。同样地,可以使用 notna()方法针对一列或所有 DataFrame;也可以使用 isna()方法,其操作与 notna()方法相反。这样,我们很容易确定 DataFrame 中什么是 NaN,什么不是 NaN。

```
>>> left = pd.DataFrame({"density": [119, 206, 240, 94],
...         "median_age": [42, 47, 46, 45],
...         "population": [65, 60, 83, 46],
...         "population_change": [0.22, -0.15, 0.32, 0.04],
...         "country": ['France', 'Italy', 'Germany', 'Spain']})
>>> right = pd.DataFrame({"world_share": [0.84, 0.78, 1.07, 0.60],
...         "population": [65, 60, 85, 46],
...         "country": ['France', 'Italy', 'Germany', 'Spain']})
>>> result = pd.merge(left, right, how="outer", on=["country", "population"])
>>> result
   density  median_age  population  population_change  country  world_share
0  119.0    42.0        65          0.22               France   0.84
1  206.0    47.0        60          -0.15              Italy    0.78
2  240.0    46.0        83          0.32               Germany  NaN
3  94.0     45.0        46          0.04               Spain    0.60
4  NaN      NaN         85          NaN                Germany  1.07
>>> result['density'].dropna()
0  119.0
1  206.0
2  240.0
3  94.0
Name: density, dtype: float64
>>> result.dropna()
   density  median_age  population  population_change  country  world_share
0  119.0    42.0        65          0.22               France   0.84
1  206.0    47.0        60          -0.15              Italy    0.78
3  94.0     45.0        46          0.04               Spain    0.60
>>> result[result['density'].notna()]
   density  median_age  population  population_change  country  world_share
0  119.0    42.0        65          0.22               France   0.84
1  206.0    47.0        60          -0.15              Italy    0.78
2  240.0    46.0        83          0.32               Germany  NaN
3  94.0     45.0        46          0.04               Spain    0.60
```

再举一个例子,我们可以使用 dropna()方法从整个 DataFrame 的一列中删除值。对于列,只删除了 NaN 值,但是在整个 DataFrame 中,删除了其中包含 NaN 的任何行。上述情况可能并不理想,相反,如果希望删除其中一列具有 NaN 的行,可以通过在 DataFrame 的列上调用 notna()方法实现。

16.5 DataFrame 方法

本节将展示一些可以应用于 DataFrame 的方法。Pandas 提供的方法很多，前面演示了 sum()方法，下面主要介绍一些更常见的数学方法。在这里，我们导入 Seaborn 库，并加载其附带的以 DataFrame 形式提供的 iris 数据集[①]。

```
>>> import seaborn as sns
>>> iris = sns.load_dataset('iris')
>>> iris.head()
   sepal_length   sepal_width   petal_length   petal_width   species
0   5.1           3.5           1.4            0.2           setosa
1   4.9           3.0           1.4            0.2           setosa
2   4.7           3.2           1.3            0.2           setosa
3   4.6           3.1           1.5            0.2           setosa
4   5.0           3.6           1.4            0.2           setosa
>>> iris.tail()
     sepal_length  sepal_width   petal_length   petal_width   species
145   6.7          3.0           5.2            2.3           virginica
146   6.3          2.5           5.0            1.9           virginica
147   6.5          3.0           5.2            2.0           virginica
148   6.2          3.4           5.4            2.3           virginica
149   5.9          3.0           5.1            1.8           virginica
>>> iris.head(10)
   sepal_length   sepal_width   petal_length   petal_width   species
0   5.1           3.5           1.4            0.2           setosa
1   4.9           3.0           1.4            0.2           setosa
2   4.7           3.2           1.3            0.2           setosa
3   4.6           3.1           1.5            0.2           setosa
4   5.0           3.6           1.4            0.2           setosa
5   5.4           3.9           1.7            0.4           setosa
6   4.6           3.4           1.4            0.3           setosa
7   5.0           3.4           1.5            0.2           setosa
8   4.4           2.9           1.4            0.2           setosa
9   4.9           3.1           1.5            0.1           setosa
>>> iris.columns
Index(['sepal_length', 'sepal_width', 'petal_length', 'petal_width',
       'species'],
      dtype = 'object')
```

---

① Seaborn 是一个基于 Python 的统计图形库，用于数据可视化和探索性数据分析。它内置了十几个示例数据集，可以通过 sns.get_dataset_names()查看数据集种类。iris 数据集也称鸢尾花数据集，是一类多重变量分析的数据集。iris 数据集包含 150 个数据样本，分为三类，每类 50 个数据，每个数据有 4 个属性。可通过花萼长度（sepal length）、花萼宽度（sepal width）、花瓣长度（petal length）和花瓣宽度（petal width）这 4 个属性预测鸢尾花属于山鸢尾（setosa）、杂色鸢尾（versicolour）和弗吉尼亚鸢尾（virginica）3 类中的哪一类。（摘自百度百科）

导入 Seaborn 库并加载 iris 数据集后,可以使用 head() 方法访问 DataFrame 的顶部,默认情况下,head() 方法会提供前五行数据。此外,可以用 tail() 方法获得后五行数据。通过将数字传递给 head() 或 tail() 方法可以得到指定行数的数据,如果只需要返回列,则可以使用 columns() 方法。

导入和访问数据后,下面演示一些可以应用的方法。

```
>>> iris.count()
sepal_length    150
sepal_width     150
petal_length    150
petal_width     150
species         150
dtype: int64
>>> iris.count().sepal_length
150
>>> iris['sepal_length'].count()
150
>>> len(iris)
150
```

从上述示例可以看出,我们可以对 DataFrame 和其中的列应用 count() 方法,当应用于 DataFrame 时返回每个列的长度。我们还可以通过使用 count() 方法,在末尾加上列名或在 iris 后面加上列名,然后应用 count() 方法来获得特定的列长度。如果想得到整个 DataFrame 中的行数,可以在 DataFrame 上使用 len() 方法。

```
>>> iris.corr()
              sepal_length  sepal_width  petal_length  petal_width
sepal_length  1.000000      -0.117570    0.871754      0.817941
sepal_width   -0.117570     1.000000     -0.428440     -0.366126
petal_length  0.871754      -0.428440    1.000000      0.962865
petal_width   0.817941      -0.366126    0.962865      1.000000
>>> iris.corr()['petal_length']['sepal_length']
0.8717537758865838
>>> iris.cov()
              sepal_length  sepal_width  petal_length  petal_width
sepal_length  0.685694      -0.042434    1.274315      0.516271
sepal_width   -0.042434     0.189979     -0.329656     -0.121639
petal_length  1.274315      -0.329656    3.116278      1.295609
petal_width   0.516271      -0.121639    1.295609      0.581006
>>> iris.cov()['sepal_length']
sepal_length    0.685694
sepal_width     -0.042434
petal_length    1.274315
petal_width     0.516271
Name: sepal_length, dtype: float64
>>> iris.cov()['sepal_length']['sepal_width']
-0.0424340044742729
```

应用于 DataFrame 的 corr() 方法提供了每个变量之间的相关性，我们可以通过传递一个列名获取该列与其他列之间的相关性，或者通过传递两个列名来获得两列之间的相关性。计算变量之间协方差的 cov() 方法也同样适用。

```
>>> iris.cumsum().head()
   sepal_length  sepal_width  petal_length  petal_width  species
0  5.1           3.5          1.4           0.2          setosa
1  10.0          6.5          2.8           0.4          setosasetosa
2  14.7          9.7          4.1           0.6          setosasetosasetosa
3  19.3          12.8         5.6           0.8          setosasetosasetosasetosa
4  24.3          16.4         7.0           1.0          setosasetosasetosasetosasetosa
>>> iris.columns
Index(['sepal_length', 'sepal_width', 'petal_length', 'petal_width',
       'species'],
      dtype = 'object')
>>> iris.cumsum()[['sepal_length', 'sepal_width', 'petal_length',
... 'petal_width']].tail()
     sepal_length    sepal_width    petal_length    petal_width
145  851.6           446.7          543.0           171.9
146  857.9           449.2          548.0           173.8
147  864.4           452.2          553.2           175.8
148  870.6           455.6          558.6           178.1
149  876.5           458.6          563.7           179.9
```

接下来介绍 cumsum() 方法，该方法提供了列数据的累积和，参见上面示例。现在，对于那些数值类型的列，该值按预期上升，当前值加上前一个值，以此类推，建立了一个递增的值。如果是一个基于字符的列时，就会出现不同情况。此时的累积值只是这些值的拼接，结果看起来很奇怪。为了便于阅读，可以通过指定要显示的列的列表来限制显示的内容。从上述示例可以看出，我们甚至可以在结尾加上 tail() 方法。

```
>>> iris.describe()
       sepal_length  sepal_width  petal_length  petal_width
count  150.000000    150.000000   150.000000    150.000000
mean   5.843333      3.057333     3.758000      1.199333
std    0.828066      0.435866     1.765298      0.762238
min    4.300000      2.000000     1.000000      0.100000
25%    5.100000      2.800000     1.600000      0.300000
50%    5.800000      3.000000     4.350000      1.300000
75%    6.400000      3.300000     5.100000      1.800000
max    7.900000      4.400000     6.900000      2.500000
>>> iris.sepal_length.describe()
count  150.000000
mean   5.843333
std    0.828066
min    4.300000
25%    5.100000
50%    5.800000
75%    6.400000
max    7.900000
Name: sepal_length, dtype: float64
```

上面示例中使用了 describe() 方法,借助该方法可以获得计数(count)、平均值(mean)、标准差(std)、最小值(min)、最大值(max)和 25%、50% 及 75% 等数值。此方法仅适用于类型为可计算数值的列,因此不包括物种(species)列。我们也可以在某列上单独使用这个方法,示例中没有使用方括号方法来访问列,而是使用了点方法,使用点和列名来访问值,然后在最后加上 describe() 方法。

```
>>> iris.max()
sepal_length       7.9
sepal_width        4.4
petal_length       6.9
petal_width        2.5
species       virginica
dtype: object
>>> iris.sepal_length.max()
7.9
```

接下来看一下 max() 方法。此例中,我们将 max() 方法应用于整个 DataFrame,得到了可以获得的每个列的最大值。另外,我们也展示了在某一列上应用该方法,这点与上一示例方式相同。

```
>>> iris.sepal_width.mean()
3.0573333333333337
>>> iris.mean(0)
sepal_length    5.843333
sepal_width     3.057333
petal_length    3.758000
petal_width     1.199333
dtype: float64
>>> iris.mean(1).head()
0    2.550
1    2.375
2    2.350
3    2.350
4    2.550
dtype: float64
>>> iris.mean(1).tail()
145    4.300
146    3.925
147    4.175
148    4.325
149    3.950
dtype: float64
```

下面要讲的方法是 mean(),其用于返回平均值。mean() 是一种常见的计算方法,经常会用到。在上面的示例中,我们首先使用点语法进行了平均值的计算,然后将该方法应用于单个列的平均值计算,但在 mean() 方法中传入了 0 或 1 的参数,该参数表示要跨列或跨行应用。还有许多不同的方法可以应用于 DataFrame，下面列出了一些十分有用的方法。

- median()：返回算术中值；
- min()：返回最小值；
- max()：返回最大值；
- mode()：返回最常出现的值；
- std()：返回标准偏差；
- sum()：返回算术和；
- var()：返回方差。

这些方法的使用示例如下所示：

```
>>> import seaborn as sns
>>> iris = sns.load_dataset('iris')
>>> iris.sepal_length.median()
5.8
>>> iris.sepal_length.min()
4.3
>>> iris.sepal_length.mode()
0    5.0
Name: sepal_length, dtype: float64
>>> iris.sepal_length.max()
7.9
>>> iris.sepal_length.std()
0.828066127977863
>>> iris.sepal_length.sum()
876.5
>>> iris.sepal_length.var()
0.6856935123042507
```

16.6 缺失值处理

本节将介绍可以跨 DataFrame 应用的方法及如何处理丢失的数据。在这里，首先按照之前的方式设置 DataFrame，并在 DataFrame 中引入一些 NaN（缺失值）条目[①]。

```
data = pd.DataFrame({"A": [1, 2.1, np.nan, 4.7, 5.6, 6.8],
        "B": [.25, np.nan, np.nan, 4, 12.2, 14.4]})
>>> import pandas as pd
>>> import numpy as np
>>>
>>> data = pd.DataFrame({"A": [1, 2.1, np.nan, 4.7, 5.6, 6.8],
```

---

① 下述示例中，原书中 data_2 未定义，在此将这两条语句注释掉了，另外原书有些结果没有展示，在此进行了补充。

```
...    "B": [.25, np.nan, np.nan, 4, 12.2, 14.4]})
>>> data
      A      B
0    1.0    0.25
1    2.1    NaN
2    NaN    NaN
3    4.7    4.00
4    5.6    12.20
5    6.8    14.40
>>> data.dropna(axis = 0)
      A      B
0    1.0    0.25
3    4.7    4.00
4    5.6    12.20
5    6.8    14.40
>>> data.dropna(axis = 1)
Empty DataFrame
Columns: []
Index: [0, 1, 2, 3, 4, 5]
>>> data.where(pd.notna(data), data.mean(), axis = "columns")
      A       B
0    1.00    0.2500
1    2.10    7.7125
2    4.04    7.7125
3    4.70    4.0000
4    5.60    12.2000
5    6.80    14.4000
>>> data.fillna(data.mean()["B":"C"])
      A       B
0    1.0     0.2500
1    2.1     7.7125
2    NaN     7.7125
3    4.7     4.0000
4    5.6     12.2000
5    6.8     14.4000
>>> data.fillna(data.mean())
      A       B
0    1.00    0.2500
1    2.10    7.7125
2    4.04    7.7125
3    4.70    4.0000
4    5.60    12.2000
5    6.80    14.4000
>>> #data_2.fillna(method = "pad")
>>> #data_2.fillna(method = "bfill")
>>> data.interpolate()
      A      B
0    1.0    0.25
1    2.1    1.50
2    3.4    2.75
```

```
3      4.7       4.00
4      5.6      12.20
5      6.8      14.40
>>> data.interpolate(method = "barycentric")
       A         B
0      1.00      0.250
1      2.10     -7.660
2      3.53     -4.515
3      4.70      4.000
4      5.60     12.200
5      6.80     14.400
>>> data.interpolate(method = "pchip")
       A          B
0      1.00000    0.250000
1      2.10000    0.672808
2      3.43454    1.928950
3      4.70000    4.000000
4      5.60000   12.200000
5      6.80000   14.400000
>>> data.interpolate(method = "akima")
       A           B
0      1.000000    0.250000
1      2.100000   -0.873316
2      3.406667    0.320034
3      4.700000    4.000000
4      5.600000   12.200000
5      6.800000   14.400000
>>> data.interpolate(method = "spline", order = 2)
       A           B
0      1.000000    0.250000
1      2.100000   -0.428598
2      3.404545    1.206900
3      4.700000    4.000000
4      5.600000   12.200000
5      6.800000   14.400000
>>> data.interpolate(method = "polynomial", order = 2)
       A           B
0      1.000000    0.250000
1      2.100000   -2.703846
2      3.451351   -1.453846
3      4.700000    4.000000
4      5.600000   12.200000
5      6.800000   14.400000
```

可以借助插值法 interpolate() 对 NaN 进行填充。interpolate() 默认执行线性插值，不带参数，此时会忽略索引，将值视为等距分布，在已有值之间进行线性填充。下面给出其他插值的简要描述。

- barycentric 插值：构造一个通过给定点集的多项式。

- pchip 插值：一维单调三次插值。
- akima 插值：给定向量 *x* 和 *y*，拟合分段三次多项式。
- spline 插值：样条线数据插值器，可以传递样条线的顺序。
- polynomial 插值：多项式数据插值器，可以传递多项式的阶数。

有关更多信息，请参阅 SciPy 文档。

接下来，我们将考虑对序列进行插值，并显示一些可以传递的可选参数。

```
>>> ser = pd.Series([np.nan, np.nan, 5, np.nan, np.nan, np.nan,
... 13, np.nan])
>>> ser
0    NaN
1    NaN
2    5.0
3    NaN
4    NaN
5    NaN
6   13.0
7    NaN
dtype: float64
>>> ser.interpolate()   ①
0    NaN
1    NaN
2    5.0
3    7.0
4    9.0
5   11.0
6   13.0
7   13.0
dtype: float64
>>> ser.interpolate(limit = 1)   ②
0    NaN
1    NaN
2    5.0
3    7.0
4    NaN
5    NaN
6   13.0
7   13.0
dtype: float64
>>> ser.interpolate(limit = 1, limit_direction = "backward")   ③
0    NaN
1    5.0
2    5.0
3    NaN
4    NaN
5   11.0
6   13.0
7    NaN
```

```
dtype: float64
>>> ser.interpolate(limit = 1, limit_direction = "both") ④
0    NaN
1    5.0
2    5.0
3    7.0
4    NaN
5    11.0
6    13.0
7    13.0
dtype: float64
>>> ser.interpolate(limit_direction = "both") ⑤
0    5.0
1    5.0
2    5.0
3    7.0
4    9.0
5    11.0
6    13.0
7    13.0
dtype: float64
>>> ser.interpolate(limit_direction = "both", limit_area = "inside",
... limit = 1) ⑥
0    NaN
1    NaN
2    5.0
3    7.0
4    NaN
5    11.0
6    13.0
7    NaN
dtype: float64
>>> ser.interpolate(limit_direction = "backward",
... limit_area = "outside") ⑦
0    5.0
1    5.0
2    5.0
3    NaN
4    NaN
5    NaN
6    13.0
7    NaN
dtype: float64
>>> ser.interpolate(limit_direction = "both",
... limit_area = "outside") ⑧
0    5.0
1    5.0
2    5.0
3    NaN
4    NaN
```

```
5    NaN
6    13.0
7    13.0
dtype: float64
```

上述代码中，示例①通过采用默认参数的interpolate()方法进行线性插值，可通过更改可选参数来研究插值效果。示例②采用了limit=1参数，表示只能对跨过任何值的一个NaN进行插值，因此在序列中仍然有NaN数据。示例③的limit参数仍然为1，又添加了limit_direction参数对方向进行限制，该方向设置为向后，这样只是在现有值的旁边插入一个值，但与以前不同的是，它是反方向。示例④将参数limit_direction设置为both，这样在向前和向后两个方向上都进行一个值的插值。示例⑤移除了limit参数，仍然设置limit_direction为both，此时可以看到所有NaN都已插值。示例⑥～⑧增加了limit_area参数，该参数有inside和outside两个选项（默认为None），当设置为inside时，NaN仅在被有效值包围时才填充，而当设置为outside时，它仅填充有效值之外的值。在这里，我们用示例展示了limit、limit_direction和limit_area参数不同取值时的执行情况。

接下来介绍replace()方法[①]。

```
>>> import seaborn as sns
>>> iris = sns.load_dataset('iris')
>>> iris.sepal_length.unique()
array([5.1, 4.9, 4.7, 4.6, 5. , 5.4, 4.4, 4.8, 4.3, 5.8, 5.7, 5.2, 5.5,
       4.5, 5.3, 7. , 6.4, 6.9, 6.5, 6.3, 6.6, 5.9, 6. , 6.1, 5.6, 6.7,
       6.2, 6.8, 7.1, 7.6, 7.3, 7.2, 7.7, 7.4, 7.9])
>>> iris.sepal_width.unique()
array([3.5, 3. , 3.2, 3.1, 3.6, 3.9, 3.4, 2.9, 3.7, 4. , 4.4, 3.8, 3.3,
       4.1, 4.2, 2.3, 2.8, 2.4, 2.7, 2. , 2.2, 2.5, 2.6])
>>> iris.petal_length.unique()
array([1.4, 1.3, 1.5, 1.7, 1.6, 1.1, 1.2, 1. , 1.9, 4.7, 4.5, 4.9, 4. ,
       4.6, 3.3, 3.9, 3.5, 4.2, 3.6, 4.4, 4.1, 4.8, 4.3, 5. , 3.8, 3.7,
       5.1, 3. , 6. , 5.9, 5.6, 5.8, 6.6, 6.3, 6.1, 5.3, 5.5, 6.7, 6.9,
       5.7, 6.4, 5.4, 5.2])
>>> iris.petal_width.unique()
array([0.2, 0.4, 0.3, 0.1, 0.5, 0.6, 1.4, 1.5, 1.3, 1.6, 1. , 1.1, 1.8,
       1.2, 1.7, 2.5, 1.9, 2.1, 2.2, 2. , 2.4, 2.3])
>>> iris.replace(2.3, 2).head()
   sepal_length  sepal_width  petal_length  petal_width  species
0  5.1           3.5          1.4           0.2          setosa
1  4.9           3.0          1.4           0.2          setosa
2  4.7           3.2          1.3           0.2          setosa
3  4.6           3.1          1.5           0.2          setosa
4  5.0           3.6          1.4           0.2          setosa
```

[①] 下述示例代码中，unique()是Pandas中Series的一个方法，返回一列中所有的唯一值，即一列中所有非重复数据，输出为array对象。

```
>>> iris.species.unique()
array(['setosa', 'versicolor', 'virginica'], dtype = object)
>>> iris.replace(['setosa', 'versicolor', 'virginica'],
... ['set', 'ver', 'vir']).head()
   sepal_length    sepal_width    petal_length    petal_width    species
0  5.1             3.5            1.4             0.2            set
1  4.9             3.0            1.4             0.2            set
2  4.7             3.2            1.3             0.2            set
3  4.6             3.1            1.5             0.2            set
4  5.0             3.6            1.4             0.2            set
>>> iris.replace(['setosa', 'versicolor', 'virginica'],['set','ver','vir'])['species'].
unique()
array(['set', 'ver', 'vir'], dtype = object)
```

16.7 数据分组、聚合

本节将介绍数据分组、聚合的概念。数据分组的功能非常强大,借助该方法,我们能够同时创建和操作数据组。示例代码如下:

```
>>> import seaborn as sns
>>> iris = sns.load_dataset('iris')
>>> groupby = iris.groupby('species')
>>> groupby.sum()
            sepal_length    sepal_width    petal_length    petal_width
species
setosa      250.3           171.4          73.1            12.3
versicolor  296.8           138.5          213.0           66.3
virginica   329.4           148.7          277.6           101.3
>>> groupby.mean()
            sepal_length    sepal_width    petal_length    petal_width
species
setosa      5.006           3.428          1.462           0.246
versicolor  5.936           2.770          4.260           1.326
virginica   6.588           2.974          5.552           2.026
```

上面示例中,我们在 iris 数据集上应用了 groupby() 方法,并根据 species 列对数据进行分组得到 groupby 对象,在该对象上可以应用 sum()、mean() 等方法进行求和、计算平均值等。这样,对于数据集所有列中 species 不同类型的数据都会应用该方法。

接下来将演示如何在分组上进行循环操作。如前所述,首先设置 DataFrame,然后在分组上循环,并在循环中打印分组的名称和内容,这样可以形象化地了解分组对数据的影响。

```
>>> groupby = iris.groupby('species')
>>> for name, group in groupby:
```

```
...        print(name)
...        print(group.head())
...
setosa
     sepal_length    sepal_width    petal_length    petal_width    species
0    5.1             3.5            1.4             0.2            setosa
1    4.9             3.0            1.4             0.2            setosa
2    4.7             3.2            1.3             0.2            setosa
3    4.6             3.1            1.5             0.2            setosa
4    5.0             3.6            1.4             0.2            setosa
versicolor
     sepal_length    sepal_width    petal_length    petal_width    species
50   7.0             3.2            4.7             1.4            versicolor
51   6.4             3.2            4.5             1.5            versicolor
52   6.9             3.1            4.9             1.5            versicolor
53   5.5             2.3            4.0             1.3            versicolor
54   6.5             2.8            4.6             1.5            versicolor
virginica
     sepal_length    sepal_width    petal_length    petal_width    species
100  6.3             3.3            6.0             2.5            virginica
101  5.8             2.7            5.1             1.9            virginica
102  7.1             3.0            5.9             2.1            virginica
103  6.3             2.9            5.6             1.8            virginica
104  6.5             3.0            5.8             2.2            virginica
```

下面介绍应用于 groupby 对象的 aggregate() 方法。与前述方法一样进行数据设置，然后将 aggregate() 方法应用于 groupby 对象，传入想要聚合的内容。在这个例子中使用了 np.sum 方法，它可以用于分组数据[①]。

```
>>> grouped = iris.groupby('species')
>>> grouped.aggregate(np.sum)
            sepal_length    sepal_width    petal_length    petal_width
species
setosa      250.3           171.4          73.1            12.3
versicolor  296.8           138.5          213.0           66.3
virginica   329.4           148.7          277.6           101.3
```

通过引入 as_index 参数可以对上面的示例进行扩展。在这里，我们使用与前面示例相同的 DataFrame，对 species 进行 groupby() 方法操作，并将 as_index 参数设置为 False。这样做的目的是创建一个关于 species 的分组，但在输出中保留 species 作为列，并使用想要分组的值。在本例中将 sum() 方法应用于分组，因此分组内的所有其他列都将被求和。

---

① 本示例运行结果与原书不同，结果采用译者运行结果。

```
>>> iris.groupby('species', as_index = False).sum()
    species     sepal_length  sepal_width  petal_length  petal_width
0   setosa      250.3         171.4        73.1          12.3
1   versicolor  296.8         138.5        213.0         66.3
2   virginica   329.4         148.7        277.6         101.3
```

对于 groupby 对象来说，还可以应用一些其他方法，这些方法也很有用，如下所示：

```
>>> grouped = iris.groupby('species')
>>> grouped.size()
species
setosa        50
versicolor    50
virginica     50
dtype: int64
>>> grouped['sepal_length'].describe()
            count  mean   std       min  25%    50%  75%  max
species
setosa      50.0   5.006  0.352490  4.3  4.800  5.0  5.2  5.8
versicolor  50.0   5.936  0.516171  4.9  5.600  5.9  6.3  7.0
virginica   50.0   6.588  0.635880  4.9  6.225  6.5  6.9  7.9
```

我们还可以对分组应用不同的方法，下面示例展示了将 NumPy 的 sum()、mean() 和 std() 等多种方法应用于分组数据。首先创建了与上一个示例相同的 DataFrame 和分组；然后使用 agg() 方法进行聚合，参数是要应用的方法列表，每个方法都会应用于数据分组；最后将 lambda 匿名函数应用到了 groupby 对象。

```
>>> grouped = iris.groupby('species')
>>> grouped['sepal_length'].agg([np.sum, np.mean, np.std])
            sum    mean   std
species
setosa      250.3  5.006  0.352490
versicolor  296.8  5.936  0.516171
virginica   329.4  6.588  0.635880
>>> grouped.agg({lambda x: np.std(x, ddof = 1)})
            sepal_length  sepal_width  petal_length  petal_width
            <lambda>      <lambda>     <lambda>      <lambda>
species
setosa      0.352490      0.379064     0.173664      0.105386
versicolor  0.516171      0.313798     0.469911      0.197753
virginica   0.635880      0.322497     0.551895      0.274650
```

接下来要展示的是通过使用 nlargest() 和 nsmalest() 方法获得一个分组中的最大值和最小值，传入的参数为整数，指示返回值的数量。下面示例中，可以看到返回了每个分组中对应数量的最大值和最小值。

```
>>> grouped = iris.groupby('species')
>>> grouped['sepal_length'].nlargest(3)
species
setosa       14     5.8
             15     5.7
             18     5.7
versicolor   50     7.0
             52     6.9
             76     6.8
virginica    131    7.9
             117    7.7
             118    7.7
Name: sepal_length, dtype: float64
>>> grouped['petal_length'].nsmallest(4)
species
setosa       22     1.0
             13     1.1
             14     1.2
             35     1.2
versicolor   98     3.0
             57     3.3
             93     3.3
             60     3.5
virginica    106    4.5
             126    4.8
             138    4.8
             121    4.9
Name: petal_length, dtype: float64
```

下面示例介绍了 apply() 方法，它是一个非常有用的应用函数的方法。首先，按照之前的方式设置数据，并按 species 列进行分组[①]。随后，在分组上使用 apply() 方法，将其中的内容应用到 groupby 对象。需要注意的是，在 DataFrame 和 Series 上也可以使用 apply() 方法。此外，本例中还将一个自定义函数 f 应用于 groupby 对象。

```
>>> grouped = iris.groupby('species', group_keys = False)
>>> def f(group):
...     return pd.DataFrame({"original" : group,
...         "demeaned" : group - group.mean()})
...
>>> grouped['petal_length'].apply(f).head()
     original    demeaned
0    1.4         -0.062
1    1.4         -0.062
2    1.3         -0.162
3    1.5          0.038
4    1.4         -0.062
```

---

① 本示例中，需要在 groupby() 方法中加入参数 group_keys=False，以便兼容以前的用法，否则会产生警告。

下面示例将给出 qcut()方法（一个很好用的 Pandas 方法），该方法根据传入的参数将数据切成大小相等的桶，实现分箱操作。这里，我们将 qcut()方法应用于 iris 数据集的 sepal_length 列，根据值为 0、0.25、0.5、0.75 和 1 的列表进行分箱，并将分箱数据赋值给 factor 变量。当把 factor 变量传给 groupby()方法时，groupby 的 mean()方法会返回每一个分箱数据的平均值，并显示分箱中的最小值和最大值。

```
>>> factor = pd.qcut(iris['sepal_length'], [0, 0.25, 0.5, 0.75, 1.0])
>>> factor.head()
0    (4.2989999999999995, 5.1]
1    (4.2989999999999995, 5.1]
2    (4.2989999999999995, 5.1]
3    (4.2989999999999995, 5.1]
4    (4.2989999999999995, 5.1]
Name: sepal_length, dtype: category
Categories(4, interval[float64, right]): [(4.2989999999999995, 5.1] <(5.1, 5.8] <(5.8, 6.4] <(6.4, 7.9]]
>>> iris.groupby(factor).mean()
                            sepal_length   sepal_width   petal_length   petal_width
sepal_length
(4.2989999999999995, 5.1]   4.856098       3.175610      1.707317       0.353659
(5.1, 5.8]                  5.558974       3.089744      3.256410       0.989744
(5.8, 6.4]                  6.188571       2.868571      4.908571       1.682857
(6.4, 7.9]                  6.971429       3.071429      5.568571       1.940000
```

到目前为止，我们已经研究了如何对单个列进行分组，下面研究如何对多个列进行分组。由于前面使用的 iris 数据集不是研究多个列分组的最佳设置，在此替换成了 tips（小费）数据集[①]。tips 数据集包含以下列：

- total_bill：消费总金额；
- tip：小费金额；
- sex：顾客性别；
- smoker：顾客是否吸烟；
- day：就餐日期，包括星期几；
- time：就餐时间；
- size：就餐人数。

考虑到有些列的局限性，最好将多个列进行分组研究，下面将按 sex 和 smoker 进行分组。

```
>>> tips = sns.load_dataset('tips')
>>> tips.head()
```

---

① tips 数据集是 Seaborn 库中自带的数据集，其中含有 7 个字段，共 244 条数据。

```
       total_bill   tip     sex     smoker   day    time     size
0      16.99        1.01    Female  No       Sun    Dinner   2
1      10.34        1.66    Male    No       Sun    Dinner   3
2      21.01        3.50    Male    No       Sun    Dinner   3
3      23.68        3.31    Male    No       Sun    Dinner   2
4      24.59        3.61    Female  No       Sun    Dinner   4
>>> grouped = tips.groupby(['sex', 'smoker'])
>>> grouped.sum()
                total_bill      tip         size
sex     smoker
Male    Yes     1337.07         183.07      150
        No      1919.75         302.00      263
Female  Yes     593.27          96.74       74
        No      977.68          149.77      140
>>> grouped = tips.groupby(['sex', 'smoker', 'time'])
>>> grouped.mean()
                        total_bill      tip         size
sex     smoker  time
Male    Yes     Lunch   17.374615       2.790769    2.153846
                Dinner  23.642553       3.123191    2.595745
        No      Lunch   18.486500       2.941500    2.500000
                Dinner  20.130130       3.158052    2.766234
Female  Yes     Lunch   17.431000       2.891000    2.300000
                Dinner  18.215652       2.949130    2.217391
        No      Lunch   15.902400       2.459600    2.520000
                Dinner  20.004138       3.044138    2.655172
```

从上述代码中可看到,当用两个或三个变量进行分组时,可通过在组内创建更多组合来增加返回值的数量。

下面将介绍 pivot_table()方法,其主要用于生成透视表。pivot_table()将变量组合并对数据进行分组的概念与 groupby 分组类似,而不同之处在于可以在此基础上在数据集上拓展分组,下面将通过示例进行演示:

```
>>> tips = sns.load_dataset('tips')
>>> tips.head()
     total_bill   tip     sex     smoker   day    time     size
0    16.99        1.01    Female  No       Sun    Dinner   2
1    10.34        1.66    Male    No       Sun    Dinner   3
2    21.01        3.50    Male    No       Sun    Dinner   3
3    23.68        3.31    Male    No       Sun    Dinner   2
4    24.59        3.61    Female  No       Sun    Dinner   4
>>> pd.pivot_table(tips, index = ["sex"])
            size         tip         total_bill
sex
Male        2.630573     3.089618    20.744076
Female      2.459770     2.833448    18.056897
>>> pd.pivot_table(tips, index = ["sex","smoker","day"])
```

```
                        size        tip        total_bill
sex    smoker day
Male   Yes    Thur     2.300000    3.058000    19.171000
              Fri      2.125000    2.741250    20.452500
              Sat      2.629630    2.879259    21.837778
              Sun      2.600000    3.521333    26.141333
       No     Thur     2.500000    2.941500    18.486500
              Fri      2.000000    2.500000    17.475000
              Sat      2.656250    3.256563    19.929063
              Sun      2.883721    3.115349    20.403256
Female Yes    Thur     2.428571    2.990000    19.218571
              Fri      2.000000    2.682857    12.654286
              Sat      2.200000    2.868667    20.266667
              Sun      2.500000    3.500000    16.540000
       No     Thur     2.480000    2.459600    16.014400
              Fri      2.500000    3.125000    19.365000
              Sat      2.307692    2.724615    19.003846
              Sun      3.071429    3.329286    20.824286
>>> pd.pivot_table(tips, index = ["sex","smoker","day"],
... values = ['tip'])
                          tip
sex    smoker day
Male   Yes    Thur       3.058000
              Fri        2.741250
              Sat        2.879259
              Sun        3.521333
       No     Thur       2.941500
              Fri        2.500000
              Sat        3.256563
              Sun        3.115349
Female Yes    Thur       2.990000
              Fri        2.682857
              Sat        2.868667
              Sun        3.500000
       No     Thur       2.459600
              Fri        3.125000
              Sat        2.724615
              Sun        3.329286
```

上述代码中，在每个例子中都使用了 tips 数据集，并设置了不同的索引值，从 sex 开始，扩展到 sex 和 smoker，然后是 sex、smoker 和 day 的组合。在每个示例中，当在默认情况下使用 pivot_table() 方法时，最终得到了 index 参数给出的所有可以取平均值的变量的平均值，因此只对数字变量有效。当不一定需要显示所有可用的变量时，可通过给 values 参数传入想要包含的列的列表来实现，如下所示：

```
>>> import numpy as np
>>> pd.pivot_table(tips, index = ["sex","smoker","day"],
```

```
... values = ['tip'],aggfunc = [np.mean,len])
                       mean     len
                       tip      tip
sex    smoker day
Male   Yes    Thur     3.058000  10
              Fri      2.741250   8
              Sat      2.879259  27
              Sun      3.521333  15
       No     Thur     2.941500  20
              Fri      2.500000   2
              Sat      3.256563  32
              Sun      3.115349  43
Female Yes    Thur     2.990000   7
              Fri      2.682857   7
              Sat      2.868667  15
              Sun      3.500000   4
       No     Thur     2.459600  25
              Fri      3.125000   2
              Sat      2.724615  13
              Sun      3.329286  14
```

默认情况下,当使用 pivot_table() 方法时,我们会得到变量的平均值,但是我们可以通过传入 aggfunc 参数来控制想要得到的结果,该参数包含想要应用于数据的函数列表。需要注意的是,我们传入的是 NumPy 中的 mean 函数和 Python 标准库中的 len 函数。

```
>>> pd.pivot_table(tips,index = ["sex","smoker"],values = ["tip"],
... columns = ["day"],aggfunc = [np.mean])
              mean
              tip
day           Thur     Fri       Sat       Sun
sex    smoker
Male   Yes    3.0580   2.741250  2.879259  3.521333
       No     2.9415   2.500000  3.256563  3.115349
Female Yes    2.9900   2.682857  2.868667  3.500000
       No     2.4596   3.125000  2.724615  3.329286
>>> pd.pivot_table(tips,index = ["sex","smoker"],values = ["tip"],
... columns = ["day"],aggfunc = [np.mean],margins = True)
              mean
              tip
day           Thur      Fri       Sat       Sun       All
sex    smoker
Male   Yes    3.058000  2.741250  2.879259  3.521333  3.051167
       No     2.941500  2.500000  3.256563  3.115349  3.113402
Female Yes    2.990000  2.682857  2.868667  3.500000  2.931515
       No     2.459600  3.125000  2.724615  3.329286  2.773519
All           2.771452  2.734737  2.993103  3.255132  2.998279
```

上述示例通过添加 margins 变量进行了扩展,该变量将给出行和列相关的总计。

16.8　Pandas 文件操作

本章中的示例使用的是 Seaborn 自带的数据集，虽然这很有用，但 Pandas 提供了很多可以读取外部文件的方法。如果回到前面利用 Python 读取和操作数据的内容，文件操作会更容易使用。Pandas 还允许将数据写回文件。为了说明怎样进行文件操作，我们将延续使用前面的 tips 数据集，并将其写入 csv 文件，然后再将写入的数据读回，如下所示：

```
>>> tips.head()
   total_bill   tip   sex     smoker  day   time    size
0  16.99        1.01  Female  No      Sun   Dinner  2
1  10.34        1.66  Male    No      Sun   Dinner  3
2  21.01        3.50  Male    No      Sun   Dinner  3
3  23.68        3.31  Male    No      Sun   Dinner  2
4  24.59        3.61  Female  No      Sun   Dinner  4
>>> tips.to_csv('myfile.csv', index=False)
>>> data = pd.read_csv('myfile.csv')
>>> data.head()
   total_bill   tip   sex     smoker  day   time    size
0  16.99        1.01  Female  No      Sun   Dinner  2
1  10.34        1.66  Male    No      Sun   Dinner  3
2  21.01        3.50  Male    No      Sun   Dinner  3
3  23.68        3.31  Male    No      Sun   Dinner  2
4  24.59        3.61  Female  No      Sun   Dinner  4
```

上述代码使用了 DataFrame 中的 to_csv() 方法将数据写入名为 "myfile.csv" 的文件中。请注意，该文件将保存在运行该命令的目录中；to_csv() 方法中 index 参数设置为 False 可以防止将 DataFrame 索引写入文件。从文件读回数据使用了 Pandas 提供的 read_csv() 方法，该方法可以获取 csv 文件内容，并将获取的内容生成一个 DataFrame。这些方法非常有用，因为不必担心文件写入或读取的过程。除了 read_csv() 方法之外，还有其他类型文件的读取方法。下面是一些更常用的文件读取方法，可以参阅 Pandas 文档查阅全部文件读取方法。

- read_excel()：读取 xls、xlsx、xlsm、xlsb、odf、ods 和 odt 文件类型。
- read_json()：读取有效的 json 字符串。

如果使用前面章节中的示例分别创建了名为 boston.json 和 boston.xlsx 的 JSON 和 Excel 文件，则可以使用以下代码将其读入 DataFrame 中：

```
>>> data.head()
   CRIM     NOX
0  0.00632  2.31
1  0.02731  7.07
```

```
2      0.02729     7.07
3      0.03237     2.18
4      0.06905     2.18
>>> file_name = '/path/to/boston.xlsx'
>>> data = pd.read_excel(file_name)
>>> data.head()
       CRIM        NOX
0      0.00632     2.31
1      0.02731     7.07
2      0.02729     7.07
3      0.03237     2.18
4      0.06905     2.18
```

正如上述代码所见，这些方法提供了将数据从通用格式加载到 DataFrame 中的简单方法。Pandas 还提供了 read_table() 方法，该方法可以用于一般分割文件，还支持数据库查询等操作，也支持读取 html 文件，但这超出了本书的范围，读者可自行去学习。

DataFrame 有几种类型的 to 方法，它们用法非常相似，支持多种不同的格式，如下所示：

- to_dict()；
- to_json()；
- to_html()；
- to_latex()；
- to_string()。

示例如下：

```
>>> import seaborn as sns
>>> tips = sns.load_dataset('tips')
>>> tips.head().to_json()
'{"total_bill":{"0":16.99,"1":10.34,"2":21.01,"3":23.68,"4":24.59},"tip":{"0":1.01,"1":1.66,"2":3.5,"3":3.31,"4":3.61},"sex":{"0":"Female","1":"Male","2":"Male","3":"Male","4":"Female"},"smoker":{"0":"No","1":"No","2":"No","3":"No","4":"No"},"day":{"0":"Sun","1":"Sun","2":"Sun","3":"Sun","4":"Sun"},"time":{"0":"Dinner","1":"Dinner","2":"Dinner","3":"Dinner","4":"Dinner"},"size":{"0":2,"1":3,"2":3,"3":2,"4":4}}'
>>> tips.head().to_dict()
{'total_bill': {0: 16.99, 1: 10.34, 2: 21.01, 3: 23.68, 4: 24.59}, 'tip': {0: 1.01, 1: 1.66, 2: 3.5, 3: 3.31, 4: 3.61}, 'sex': {0: 'Female', 1: 'Male', 2: 'Male', 3: 'Male', 4: 'Female'}, 'smoker': {0: 'No', 1: 'No', 2: 'No', 3: 'No', 4: 'No'}, 'day': {0: 'Sun', 1: 'Sun', 2: 'Sun', 3: 'Sun', 4: 'Sun'}, 'time': {0: 'Dinner', 1: 'Dinner', 2: 'Dinner', 3: 'Dinner', 4: 'Dinner'}, 'size': {0: 2, 1: 3, 2: 3, 3: 2, 4: 4}}
>>> tips.head().to_html()
```

```
'< table border = "1" class = "dataframe">\n< thead >\n    < tr style = "text - align: right;">\n
< th ></ th >\n      < th > total_bill </ th >\n      < th > tip </ th >\n      < th > sex </ th >\n
< th > smoker </ th >\n      < th > day </ th >\n      < th > time </ th >\n      < th > size </ th >\n
</ tr >\n </ thead >\n < tbody >\n    < tr >\n      < th > 0 </ th >\n      < td > 16.99 </ td >\n
< td > 1.01 </ td >\n      < td > Female </ td >\n      < td > No </ td >\n      < td > Sun </ td >\n
< td > Dinner </ td >\n      < td > 2 </ td >\n    </ tr >\n    < tr >\n
< th > 1 </ th >\n      < td > 10.34 </ td >\n      < td > 1.66 </ td >\n      < td > Male </ td >\n
< td > No </ td >\n      < td > Sun </ td >\n      < td > Dinner </ td >\n      < td > 3 </ td >\n
</ tr >\n    < tr >\n      < th > 2 </ th >\n      < td > 21.01 </ td >\n
< td > 3.50 </ td >\n      < td > Male </ td >\n      < td > No </ td >\n      < td > Sun </ td >\n
< td > Dinner </ td >\n      < td > 3 </ td >\n    </ tr >\n    < tr >\n      < th > 3 </ th >\n
< td > 23.68 </ td >\n      < td > 3.31 </ td >\n      < td > Male </ td >\n      < td > No </ td >\n
< td > Sun </ td >\n      < td > Dinner </ td >\n      < td > 2 </ td >\n    </ tr >\n
< tr >\n      < th > 4 </ th >\n      < td > 24.59 </ td >\n      < td > 3.61 </ td >\n
< td > Female </ td >\n      < td > No </ td >\n      < td > Sun </ td >\n      < td > Dinner </ td >\n
< td > 4 </ td >\n    </ tr >\n </ tbody >\n </ table >'
>>> tips.head().to_latex()
'\\begin{tabular}{lrrlllr}\n\\toprule\n{} & total\\_bill & tip &     sex & smoker & day &
time & size \\\\\n\\midrule\n0 &      16.99 & 1.01 & Female &     No & Sun & Dinner &     2 \\\\\n1
&      10.34 & 1.66 &   Male &     No & Sun & Dinner &     3 \\\\\n2 &      21.01 & 3.50 &   Male &     No &
Sun & Dinner &     3 \\\\\n3 &      23.68 & 3.31 &   Male &     No & Sun & Dinner &     2 \\\\\n4 &
24.59 & 3.61 & Female &     No & Sun & Dinner &     4 \\\\\n\\bottomrule\n\\end{tabular}\n'
>>> tips.head().to_string()
'   total_bill   tip     sex smoker  day    time  size\n0       16.99  1.01  Female     No  Sun  Dinner     2\n1
10.34  1.66    Male     No  Sun  Dinner     3\n2       21.01  3.50    Male     No  Sun  Dinner     3\n3    23.68
3.31    Male     No  Sun  Dinner     2\n4       24.59  3.61  Female     No  Sun  Dinner     4'
>>>
```

我们还可以使用其中一些方法将数据直接写入文件,主要支持格式如下所示:

```
>>> import seaborn as sns
>>> tips = sns.load_dataset('tips')
>>> tips.head().to_json('tips.json')
>>> tips.to_html('tips.html')
>>> tips.to_latex('tips.latex')
>>> tips.to_latex('tips.tex')
```

这些方法非常有用,对于格式正确的数据来说,可以方便地将数据读入 Pandas,或者从 Pandas 中导出数据。

本章小结

本章主要介绍了 Pandas 用于数据分析的一些高级方法。本章展示了 Pandas 是如何像在数据库中一样操作数据的,它允许我们以各种方式对数据进行连接(join)、合并(merge)、分组(group)及透视(pivot)操作。本章还介绍了 Pandas 的一些内置方法,并展示了如何处理数据的缺失值。本章所介绍的示例实际上十分简单,但 Pandas 功能强大,可以处理大型数据集,这使它成为一个非常强大的 Python 包。还值得注意的是,Pandas 与许多其他 Python 包配合很好,因此学好 Pandas 对 Python 程序员来说是至关重要的。

CHAPTER 17

第 17 章 数据可视化

对于任何数据分析而言，数据可视化都十分重要。Python 提供了大量的可视化工具。本章将详细介绍数据可视化相关内容，首先从 Pandas 创建基本绘图开始，然后讲解如何使用 Matplotlib 进行更多复杂绘图控制，最后介绍数据可视化工具 Seaborn 的使用。本章完全按照案例驱动形式展开，提供了详细的案例代码以及代码生成的图形。这样做的目的是能够直观看到由代码绘制图形的内容，同时也可以作为数据可视化的参考。本章中的示例基于 Python 中的数据集，其中许多是从文档中提取的，目的是加深对案例的理解，了解代码编写的原理。本章将会用到以下包：

```
>>> import matplotlib.pyplot as plt
>>> import numpy as np
>>> import pandas as pd
>>> import matplotlib
>>> import seaborn as sns
```

17.1 Pandas

首先从 Pandas 提供的 Series 和 DataFrame 对象研究数据可视化方法。Pandas 最大的好处是它内置了绘图方法，可以调用这些方法并生成绘图，从而可以快速、直观地呈现想要分析的数据集。示例代码和相关显示（见图 17.1）如下：

```
>>> iris = sns.load_dataset('iris')
>>> iris.sepal_length.plot()
<AxesSubplot: >
>>> plt.show()
>>> iris.sepal_length.plot()
<AxesSubplot: >
>>> plt.savefig('/path/to/file/file_name.pdf')
```

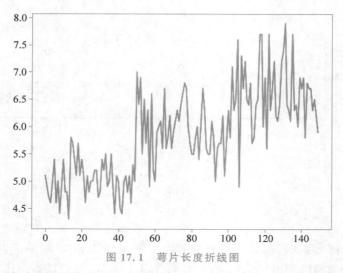

图 17.1 萼片长度折线图

上述代码和相关绘图展示了如何利用 Series 数据绘制简单的折线图。示例中给出了两种图形绘制方法,一种是利用 plt.show()直接进行图形显示,另一种利用 plt.savefig()将图形保存到文件。当需要保存图形时,需要提供要保存图形的文件路径和名称。

```
>>> plt.clf()
>>> iris.sepal_length.hist()
<AxesSubplot:>
```

上述代码使用了 hist()方法绘制直方图,与前面示例中使用的绘图方法相同,绘图结果如图 17.2 所示。需要注意的是 plt.clf()放在第 1 行,主要是为了清除之前绘制的图形,以利于新绘制图形的显示。

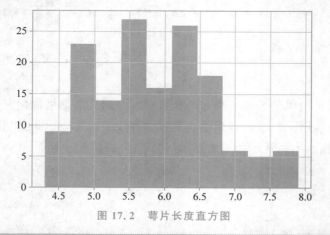

图 17.2 萼片长度直方图

```
>>> plt.clf()
>>> iris.sepal_length.plot.box()
```

```
< AxesSubplot: >
```

```
>>> plt.clf()
>>> iris.sepal_length.plot.density()
< AxesSubplot: ylabel = 'Density'>
>>> plt.show()
```

上述两段程序分别展示了如何绘制箱线图（boxplot）和密度图（densityplot），结果分别如图17.3和图17.4所示。在这两个示例中，我们使用plot方法来访问想要创建的绘图。这与之前的用法略有不同，通过查看plot方法，可以弄清创建绘图的类型。

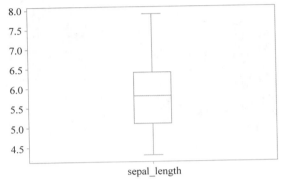

图 17.3　萼片长度箱线图

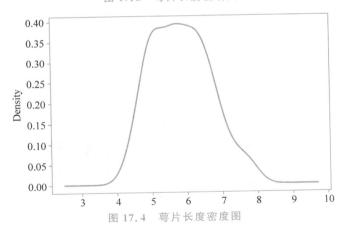

图 17.4　萼片长度密度图

```
>>> dir(iris.sepal_length.plot)
['__annotations__', '__call__', '__class__', '__delattr__', '__dict__', '__dir__', '__doc__', '_
_eq__', '__format__', '__ge__', '__getattribute__', '__gt__', '__hash__', '__init__', '__init_
subclass__', '__le__', '__lt__', '__module__', '__ne__', '__new__', '__reduce__', '__reduce_
ex__', '__repr__', '__setattr__', '__sizeof__', '__str__', '__subclasshook__', '__weakref_
', '_accessors', '_all_kinds', '_common_kinds', '_constructor', '_dataframe_kinds', '_dir_
additions', '_dir_deletions', '_get_call_args', '_hidden_attrs', '_kind_aliases', '_parent',
'_reset_cache', '_series_kinds', 'area', 'bar', 'barh', 'box', 'density', 'hexbin', 'hist', 'kde',
'line', 'pie', 'scatter']
```

通过查看 iris.sepal_length.plot 包含的方法可以看到，除了箱线图和密度图之外，还可以绘制下述类型的图形：

- 面积图(area plot)，如图 17.5 所示。
- 直方图(hist plot)，如图 17.6 所示。
- 核密度估计图(KDE plot)，如图 17.7 所示。
- 折线图(line plot)，如图 17.8 所示。

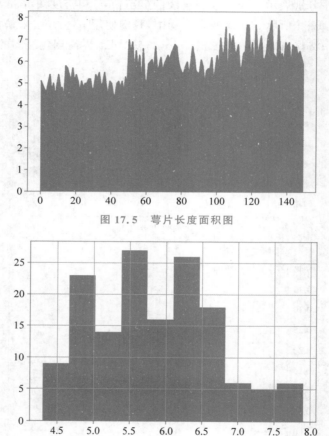

图 17.5　萼片长度面积图

图 17.6　萼片长度直方图

到目前为止，我们已经研究了应用 Series 序列数据进行图形绘制的方法，此类方法也可以应用于 DataFrame 对象的单列数据。我们也可以在整个 DataFrame 上进行绘图，与上述处理 iris 数据集的 DataFrame 对象的单列数据(sepal_length)不同，需要将 DataFrame 作为一个整体进行处理。

下面介绍如图 17.9 所示的箱线图。当将 DataFrame 作为一个整体进行绘图时，我们可以看到每个变量都有对应的箱线图，并且 Pandas 还根据 DataFrame 名称标记相应变量。

第17章 数据可视化

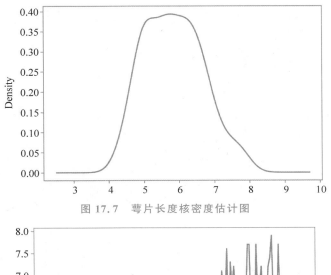

图 17.7 萼片长度核密度估计图

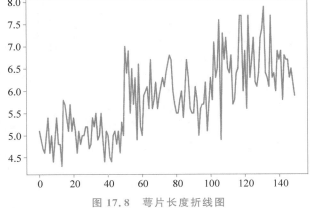

图 17.8 萼片长度折线图

```
>>> plt.clf()
>>> iris.plot.box()
<AxesSubplot: >
```

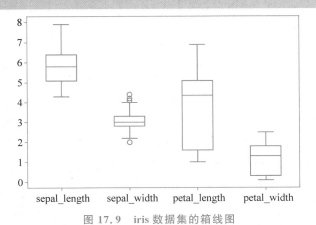

图 17.9 iris 数据集的箱线图

```
>>> plt.clf()
>>> iris.plot.density()
<AxesSubplot: ylabel = 'Density'>
>>> plt.show()
```

接下来将展示 DataFrame 的密度图和折线图,分别如图 17.10 和图 17.11 所示。与箱线图示例一样,可为每个可以获取的变量绘制一条线。箱线图的每个箱体都有标签,同样,密度图和折线图的每条线都有一个图例,每种颜色表示所指的内容。

```
>>> plt.clf()
>>> iris.plot.line()
<AxesSubplot: >
>>> plt.show()
```

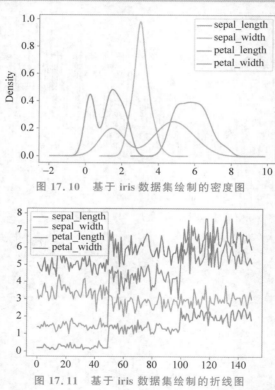

图 17.10 基于 iris 数据集绘制的密度图

图 17.11 基于 iris 数据集绘制的折线图

在最后一个内置的绘图示例中,我们将介绍 DataFrame 的散点图(scatter plot),如图 17.12 所示。与前面给出的绘图形式不同,散点图需要 x 和 y 两个变量,绘制 x 与 y 的关系图。因此,我们从 DataFrame 中传递特定列的名称,scatter()方法可以获取所需的数据,以给出 sepal_length 与 sepal_width 的散点图。

```
>>> plt.clf()
>>> iris.plot.scatter(x = 'sepal_length', y = 'sepal_width')
```

```
< AxesSubplot: xlabel = 'sepal_length', ylabel = 'sepal_width'>
>>> plt.show()
```

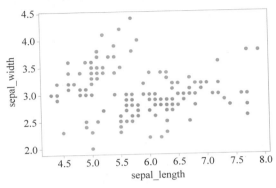

图 17.12　利用 DataFrame 数据绘制散点图

与之前一样，对于 DataFrame 还可以绘制下述类型的图形：
- 面积图，如图 17.13 所示。
- 直方图，如图 17.14 所示。
- 核密度估计图，如图 17.15 所示。

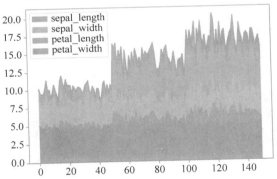

图 17.13　利用 iris DataFrame 数据绘制面积图

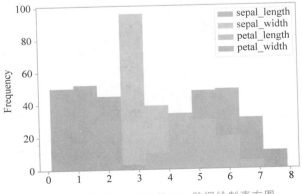

图 17.14　利用 iris DataFrame 数据绘制直方图

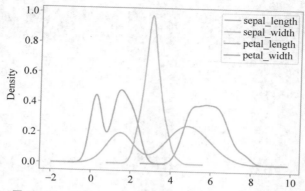

图 17.15　利用 iris DataFrame 数据绘制核密度估计图

这些示例的绘图如下所示，它们都是使用前面给出的 iris.plot 方法绘制。

有几个图没有包含在前面的示例中，即垂直条形图（bar plot）、水平条形图（barh plot）和饼图（pie plot）。饼图和水平条形图的示例分别如图 17.16 和图 17.17 所示，分别对应的代码如下所示①。这类绘图要求数据采用适合绘图的格式，因此需要对数据进行预处理为所需格式。到目前为止，所有其他绘图都是直接在 DataFrame 上进行，这里将对 tips 数据集的 DataFrame 对象进行组合，以获得可以使用的格式。

```
>>> import seaborn as sns
>>> tips = sns.load_dataset('tips')
>>> grouped = tips.groupby(['day'])
>>> grouped.sum().plot.pie(y = 'size')
```

```
>>> import seaborn as sns
>>> tips = sns.load_dataset('tips')
>>> grouped = tips.groupby(['day'])
>>> grouped.sum().plot.barh()
```

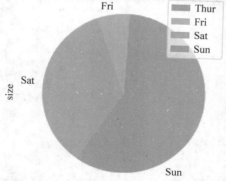

图 17.16　基于 tips 数据集的 day 和 size 数据绘制的饼图

---

① 原书所附代码有误，此处对代码进行了更正。

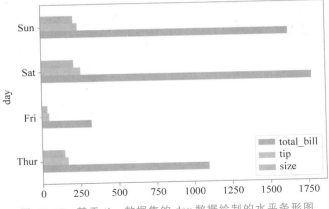

图 17.17 基于 tips 数据集的 day 数据绘制的水平条形图

17.2 Matplotlib

本节将展示一些利用 Pandas 和 Matplotlib 轻松绘制的图形,接下来将研究如何进行定制绘图。

```
>>> fig = plt.figure()
>>> plt.plot(iris.sepal_length, '-')
[<matplotlib.lines.Line2D object at 0x000001E5579F4610>]
>>> plt.plot(iris.petal_length, '--')
[<matplotlib.lines.Line2D object at 0x000001E5579F4400>]
>>> fig.savefig('iris_plot.pdf')
```

在上述代码中,我们使用了另一种方法来绘制图形,绘制结果如图 17.18 所示。从 Matplotlib 的角度来看,这种方法更为常见。本例中,使用 plot 方法绘制折线图,并通过一个额外的参数来指定线条类型,以确定是实线还是虚线。最后,使用 savefig() 方法保存绘图。我们可以通过运行以下命令找到支持的文件类型。

```
>>> fig.canvas.get_supported_filetypes()
{'eps': 'Encapsulated Postscript',
 'jpg': 'Joint Photographic Experts Group',
 'jpeg': 'Joint Photographic Experts Group',
 'pdf': 'Portable Document Format',
 'pgf': 'PGF code for LaTeX',
 'png': 'Portable Network Graphics',
 'ps': 'Postscript',
 'raw': 'Raw RGBA bitmap',
 'rgba': 'Raw RGBA bitmap',
 'svg': 'Scalable Vector Graphics',
 'svgz': 'Scalable Vector Graphics',
```

```
'tif': 'Tagged Image File Format',
'tiff': 'Tagged Image File Format',
'webp': 'WebP Image Format'}
```

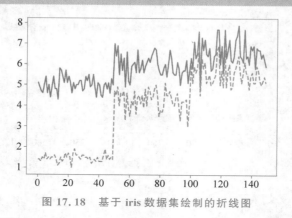

图 17.18 基于 iris 数据集绘制的折线图

下面的例子是绘制组合图（panel plot），如图 17.19 所示。绘制组合图是在一张图上绘制两个图形，而不是在一个图形上绘制两条线。

```
>>> plt.clf()
>>> plt.figure()
<Figure size 640x480 with 0 Axes>
>>> plt.subplot(2, 1, 1)
<AxesSubplot: >
>>> plt.plot(iris.sepal_length)
[<matplotlib.lines.Line2D object at 0x000001E5548EA3B0>]
>>> plt.subplot(2, 1, 2)
<AxesSubplot: >
>>> plt.plot(iris.petal_length)
[<matplotlib.lines.Line2D object at 0x000001E5549222F0>]
```

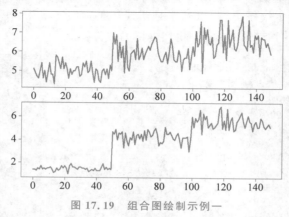

图 17.19 组合图绘制示例一

要绘制组合图，首先以通常方式创建图形，然后使用 subplot() 方法创建两个面板。

subplot()方法的参数如下：
- 行数(number of rows)；
- 列数(number of columns)；
- 索引(index)。

此例组合图中有两行一列，首先绘制第一条 sepal_length 线，接着在第二个面板中绘制第二条 petal_length 线，这样就得到了面板图的示例，如图 17.20 所示。

```
>>> plt.clf()
>>> fig, ax = plt.subplots(2)
>>> ax[0].plot(iris.sepal_length)
[<matplotlib.lines.Line2D object at 0x000001E555BF73D0>]
>>> ax[1].plot(iris.petal_length)
[<matplotlib.lines.Line2D object at 0x000001E555BF6500>]
```

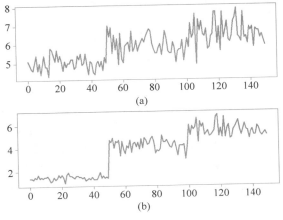

图 17.20　组合图绘制示例二

借助 subplot()方法直接设置两个子图，使用 ax 变量控制传入每个子图的内容，这样可以实现相同的效果。此时，可以直接在第一个子图中使用 plot()方法绘制萼片长度(sepal length)，在第二个子图中绘制花瓣长度(petal length)。

接下来，就可以看一下绘图效果。在 Matplotlib 中，我们可以控制绘图的样式，通过运行下面的命令，可以得到 Matplotlib 中可用的样式列表：

```
>>> plt.style.available
['Solarize_Light2', '_classic_test_patch', '_mpl-gallery', '_mpl-gallery-nogrid', 'bmh', 'classic', 'dark_background', 'fast', 'fivethirtyeight', 'ggplot', 'grayscale', 'seaborn-v0_8', 'seaborn-v0_8-bright', 'seaborn-v0_8-colorblind', 'seaborn-v0_8-dark', 'seaborn-v0_8-dark-palette', 'seaborn-v0_8-darkgrid', 'seaborn-v0_8-deep', 'seaborn-v0_8-muted', 'seaborn-v0_8-notebook', 'seaborn-v0_8-paper', 'seaborn-v0_8-pastel', 'seaborn-v0_8-poster', 'seaborn-v0_8-talk', 'seaborn-v0_8-ticks', 'seaborn-v0_8-white', 'seaborn-v0_8-whitegrid', 'tableau-colorblind10']
```

要更详细地了解这些样式的差异,可以参阅 Matplotlib 中关于样式的文档,文档中有与特定样式相关的图片。需要注意的是,我们可以通过使用不同的图表样式来改变绘图的外观。

```
>>> plt.clf()
>>> plt.style.use('seaborn-whitegrid')
>>> fig = plt.figure()
>>> ax = plt.axes()
>>> plt.plot(tips.tip, color = 'blue')
[<matplotlib.lines.Line2D object at 0x000001E555C6C040>]
>>> plt.plot(tips.tip + 5, color = 'g')
[<matplotlib.lines.Line2D object at 0x000001E555C6C1F0>]
>>> plt.plot(tips.tip + 10, color = '0.75')
[<matplotlib.lines.Line2D object at 0x000001E555C6C1C0>]
>>> plt.plot(tips.tip + 15, color = '#FFDD44')
[<matplotlib.lines.Line2D object at 0x000001E555C6C5B0>]
>>> plt.plot(tips.tip + 20, color = (1.0,0.2,0.3))
[<matplotlib.lines.Line2D object at 0x000001E555C6CDC0>]
>>> plt.plot(tips.tip + 25, color = 'chartreuse')
[<matplotlib.lines.Line2D object at 0x000001E555C6CA00>]
```

上述代码示例中,采用了 seaborn-whitegrid 绘图样式,并借助 figure() 和 axes() 方法设置图(figure)和坐标系(axes)。接下来,我们使用 plot() 方法绘制多个相同的 tips 折线图,但在每个示例中,我们将每个值加上 5,以便可以看到每条线之间的差异。示例中,每条线都使用 color 参数赋予不同的颜色。第一条线的 color 参数设置为蓝色,我们可以使用颜色作为线的名称。第二条线的 color 参数设置为单个字母"g",表示绿色。第三条线的颜色使用 0~1 的灰度值进行设置。第四条线显示了如何使用十六进制颜色编码来设置颜色。第五条线展示了如何使用 RGB 元组设置颜色,其中每个值范围为 0~1,第一个值表示红色,第二个值表示绿色,第三个值表示蓝色。结果如图 17.21 所示。

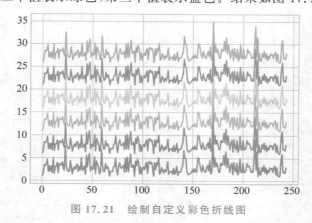

图 17.21　绘制自定义彩色折线图

接下来介绍在 Matplotlib 中如何绘制不同的线条样式,可以使用 linestyle 参数传递线条样式进行绘制,绘制结果如图 17.22 所示。

```
>>> plt.clf()
>>> plt.style.use('fivethirtyeight')
>>> fig = plt.figure()
>>> ax = plt.axes()
>>> plt.plot(tips.tip, linestyle = 'solid')
[< matplotlib.lines.Line2D object at 0x000001E555C408E0 >]
>>> plt.plot(tips.tip + 5, linestyle = 'dashed')
[< matplotlib.lines.Line2D object at 0x000001E555C8D6F0 >]
>>> plt.plot(tips.tip + 10, linestyle = 'dashdot')
[< matplotlib.lines.Line2D object at 0x000001E555BF4EB0 >]
>>> plt.plot(tips.tip + 15, linestyle = 'dotted');
[< matplotlib.lines.Line2D object at 0x000001E555BF4AC0 >]
>>> plt.plot(tips.tip + 20, linestyle = '-')
[< matplotlib.lines.Line2D object at 0x000001E555CAC6A0 >]
>>> plt.plot(tips.tip + 25, linestyle = '--')
[< matplotlib.lines.Line2D object at 0x000001E555CAD300 >]
>>> plt.plot(tips.tip + 30, linestyle = '-.')
[< matplotlib.lines.Line2D object at 0x000001E555C40730 >]
>>> plt.plot(tips.tip + 35, linestyle = ':')
[< matplotlib.lines.Line2D object at 0x000001E555CAD540 >]
```

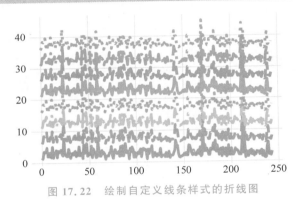

图 17.22　绘制自定义线条样式的折线图

我们也可以将线条样式和颜色组合到一个参数中来绘图不同颜色和样式的线条。例如,可以将短线样式与单字母颜色相结合进行绘制,其结果如图 17.23 所示。

```
>>> plt.clf()
>>> plt.style.use('ggplot')
>>> fig = plt.figure()
>>> ax = plt.axes()
>>> plt.plot(tips.tip, '-g') # solid green
[< matplotlib.lines.Line2D object at 0x000001E555F9E590 >]
>>> plt.plot(tips.tip + 5, '--c') # dashed cyan
[< matplotlib.lines.Line2D object at 0x000001E555F9E410 >]
>>> plt.plot(tips.tip + 10, '-.k') # dashdot black
[< matplotlib.lines.Line2D object at 0x000001E555F9E710 >]
>>> plt.plot(tips.tip + 15, ':r') # dotted red
[< matplotlib.lines.Line2D object at 0x000001E555F9E9B0 >]
```

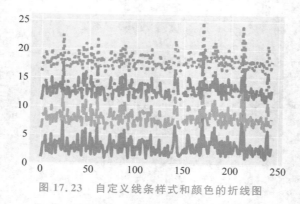

图 17.23　自定义线条样式和颜色的折线图

接下来，我们将了解如何更改坐标轴刻度范围（limits），并在绘图中添加标签（label），下面的示例展示了如何设置绘图的坐标刻度范围。

```
>>> fig = plt.figure()
>>> plt.plot(tips.tip)
[<matplotlib.lines.Line2D object at 0x00000244D16069E0>]
>>> plt.xlim(50, 200)
(50.0, 200.0)
>>> plt.ylim(0, 10)
(0.0, 10.0)
>>> fig = plt.figure()
>>> plt.plot(tips.tip)
[<matplotlib.lines.Line2D object at 0x00000244D1666C20>]
>>> plt.xlim(200, 50)
(200.0, 50.0)
>>> plt.ylim(10, 0)
(10.0, 0.0)
```

上述代码中绘制的第一个图形如图 17.24 所示，采用 tips 数据集绘制，利用 xlim() 和 ylim() 方法分别设置了 x 轴和 y 轴坐标范围。在使用这两个方法时，只需要将希望的坐标轴范围的起始值和结束值输入到 xlim() 和 ylim() 方法的开始和结束参数中，本例中 x 轴的范围是 50~200，y 轴的范围为 0~10。

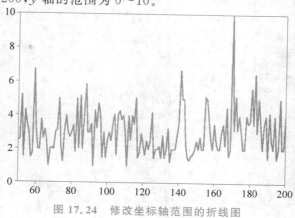

图 17.24　修改坐标轴范围的折线图

在图 17.25 中，将坐标轴的范围进行了简单的变化，分别将 x 轴的范围调整为 200～50，将 y 轴的范围调整为 10～0。

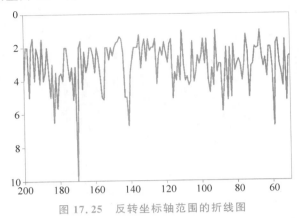

图 17.25　反转坐标轴范围的折线图

```
>>> fig = plt.figure()
>>> plt.plot(tips.tip)
[<matplotlib.lines.Line2D object at 0x00000244D4DDED40>]
>>> plt.title("Tips from the seaborn tips dataset")
Text(0.5, 1.0, 'Tips from the seaborn tips dataset')
>>> plt.xlabel("Indiviual")
Text(0.5, 0, 'Indiviual')
>>> plt.ylabel("Tip ( £ )")
Text(0, 0.5, 'Tip ( £ )')

>>> fig = plt.figure()
>>> plt.plot(tips.tip, '-g', label = 'tip')
[<matplotlib.lines.Line2D object at 0x00000244D4E32E60>]
>>> plt.plot(tips.total_bill, '-b', label = 'total bill')
[<matplotlib.lines.Line2D object at 0x00000244D4E33220>]
>>> plt.axis('equal')
(-12.15, 255.15, -1.4905000000000004, 53.3005)
>>> plt.legend()
<matplotlib.legend.Legend object at 0x00000244D4E33160>
```

上述代码第一个示例绘图结果如图 17.26 所示，使用 title()、xlabel() 和 ylabel() 方法标记了绘图标题和坐标轴名称，分别对应绘图标题、x 轴名称和 y 轴名称。我们还可以使用 legend() 方法向图中添加图例，如图 17.27 所示。添加图例时，需要在 plot() 方法中使用 label 参数指定要在图例中显示的名称。

前面展示了 Matplotlib 标准的绘图方法，接下来看一下 Matplotlib 还能绘制什么样的图形。

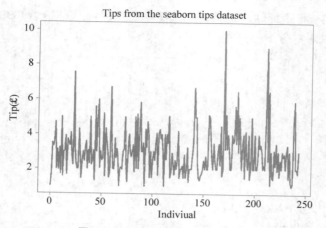

图 17.26 在折线图中添加标签

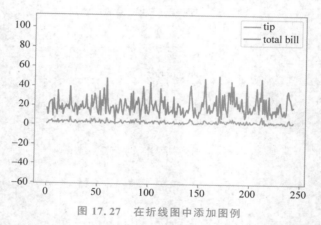

图 17.27 在折线图中添加图例

```
>>> fig = plt.figure()
>>> rng = np.random.RandomState(0)
>>> for marker in ['o', '.', ',', 'x', '+', 'v', '^', '<', '>','s', 'd']:
...     plt.plot(rng.rand(5), rng.rand(5), marker,
...         label = "marker = '{0}'".format(marker))
...
[<matplotlib.lines.Line2D object at 0x00000244D43DC7C0>]
[<matplotlib.lines.Line2D object at 0x00000244D43DCAC0>]
[<matplotlib.lines.Line2D object at 0x00000244D43DCD60>]
[<matplotlib.lines.Line2D object at 0x00000244D43DD030>]
[<matplotlib.lines.Line2D object at 0x00000244D43DD2D0>]
[<matplotlib.lines.Line2D object at 0x00000244D43DD570>]
[<matplotlib.lines.Line2D object at 0x00000244D43DD810>]
[<matplotlib.lines.Line2D object at 0x00000244D43DDAB0>]
[<matplotlib.lines.Line2D object at 0x00000244D43DDD50>]
[<matplotlib.lines.Line2D object at 0x00000244D43DDFF0>]
[<matplotlib.lines.Line2D object at 0x00000244D43DE290>]
```

```
>>> plt.legend(numpoints = 1)
<matplotlib.legend.Legend object at 0x00000244D4E78490>
>>> plt.xlim(0, 1.8)
(0.0, 1.8)
```

上述代码运行结果如图17.28所示，图中展示了如何在绘图上使用不同的标记(marker)。首先使用NumPy库的RandomState()方法生成一个随机数数组，然后遍历要绘制的标记列表，每个标记绘制5个点，数值为由随机数生成的数组，并添加一个图例显示图中每个点所代表的内容。

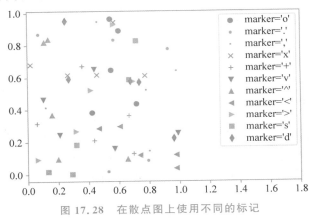

图17.28 在散点图上使用不同的标记

```
>>> fig = plt.figure()
>>> rng = np.random.RandomState(0)
>>> x = rng.randn(100)
>>> y = rng.randn(100)
>>> colors = rng.rand(100)
>>> sizes = 1000 * rng.rand(100)
>>> plt.scatter(x, y, c = colors, s = sizes, alpha = 0.3, cmap = 'viridis')
<matplotlib.collections.PathCollection object at 0x00000244D445E0B0>
>>> plt.colorbar()
<matplotlib.colorbar.Colorbar object at 0x00000244D445E440>
```

下面给出的是散点图，如图17.29所示，其中每个元素都有与其对应的颜色和大小。实现代码如上所示，首先以通常方式绘制一个图形，然后使用RandomState()生成随机数，生成了100个随机数来计算散点图中的 x 和 y 坐标。接下来，再生成100个随机数来确定颜色。最后，通过将另外100个随机数乘以1000来生成散点尺寸，使小球大小差异足够大，以便观察。得到这些数据之后，就可以将这些值传递给scatter()方法，并将颜色数组传递给c参数，将散点尺寸传递给参数s，通过alpha参数设置绘图的不透明程度，使用cmap参数设置要使用的颜色映射(colormap)。一旦绘制了散点，便可以添加颜色渐变条(colorbar)，以显示颜色相对于颜色数组中随机数所指的颜色。

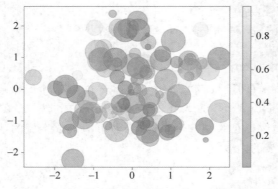

图 17.29 修改散点图中散点的尺寸和颜色

17.3 Seaborn

前面的章节介绍了利用 Pandas 对象直接绘图以及使用 Matplotlib 创建和定制绘图,本节将介绍 Seaborn 库,它是对 Matplotlib 进行二次封装而成,可以说是 Matplotlib 库的升级版,能够高度兼容 Pandas 的 DataFrame 数据结构。引述 Seaborn 官网说法,它提供的功能主要如下:

- 面向数据集的应用程序编程接口(API),用于检查多个变量之间的关系。
- 支持使用分类变量来显示观察结果或汇总统计数据。
- 可选单变量或双变量分布可视化,支持数据子集比较。
- 各类因变量的线性回归模型的自动估计和绘图。
- 方便查看复杂数据集的整体结构。
- 具有构建多图块网格的高级抽象,可以轻松构建复杂可视化视图。
- 通过几个内置主题对 Matplotlib 图形样式进行简洁控制。
- 选择调色板工具,能如实表示用户的数据模式。

以上功能说明,Seaborn 不仅可以绘制箱线图这种简单形式,而且能够实现更加简易实用的绘图,可以与 DataFrame 一起,实现更高层次的绘图。

```
>>> import matplotlib.pyplot as plt
>>> import seaborn as sns
>>> sns.set(style = "darkgrid")
>>> tips = sns.load_dataset("tips")
>>> tips.columns
Index(['total_bill', 'tip', 'sex', 'smoker', 'day', 'time', 'size'], dtype = 'object')
>>> sns.relplot(x = "total_bill", y = "tip", data = tips)
< seaborn.axisgrid.FacetGrid object at 0x00000244D44A7D60 >
```

上述代码中,依然如前所示加载 tips 数据集,然后使用 relplot()方法,并将

DataFrame 的名称作为参数传递给 data 选项,并将 x 和 y 设置为 DataFrame 中的列名。绘制结果如图 17.30 所示,绘制图形是 x 和 y 变量的散点图。

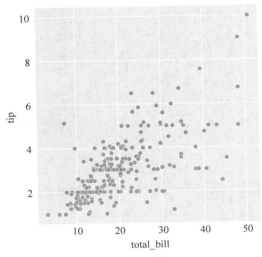

图 17.30　使用 Seaborn 库中的 relplot( )方法绘制散点图

我们可以通过添加第三个维度对其进行扩展。在这里,我们希望在相同的散点图上根据顾客是否吸烟对数据进行区分。为此,需要将 DataFrame 中名称为 smoker 的列传入 hue 选项。需要注意的是,smoker 的值为"Yes"和"No",这与我们在图 17.31 中看到的相似。

```
>>> tips.head()
   total_bill   tip    sex     smoker  day   time    size
0  16.99        1.01   Female  No      Sun   Dinner  2
1  10.34        1.66   Male    No      Sun   Dinner  3
2  21.01        3.50   Male    No      Sun   Dinner  3
3  23.68        3.31   Male    No      Sun   Dinner  2
4  24.59        3.61   Female  No      Sun   Dinner  4
>>> tips.smoker.unique()
['No', 'Yes']
Categories (2, object): ['Yes', 'No']
>>> sns.relplot(x = "total_bill", y = "tip", hue = "smoker", data = tips);
< seaborn.axisgrid.FacetGrid object at 0x00000244D1231BA0 >
```

上述代码的绘图结果可以通过颜色显示消费者是否吸烟,如图 17.31 所示。

另外,我们还可以通过设置 style 参数进行扩展,将 style 参数设置为 smoker,这将改变 smoker 中不同类别的绘图样式,本例中两种样式分别为"Yes"和"No",绘图效果如图 17.32 所示。

```
>>> sns.relplot(x = "total_bill", y = "tip", hue = "smoker",
... style = "smoker", data = tips);
< seaborn.axisgrid.FacetGrid object at 0x00000244D47A0850 >
```

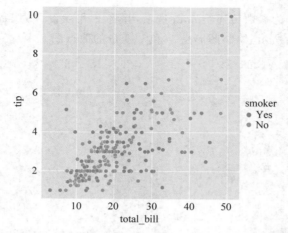

图 17.31　使用 Seaborn 库中的 relplot( )方法绘制具有第三维度数据的散点图

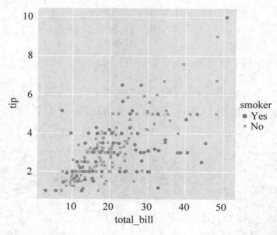

图 17.32　使用 Seaborn 库中的 relplot( )方法绘制具有第三维度数据和样式的散点图

```
>>> tips.dtypes
total_bill      float64
tip             float64
sex            category
smoker         category
day            category
time           category
size              int64
dtype: object
>>> tips.head()
   total_bill   tip     sex     smoker  day   time     size
0    16.99    1.01   Female    No     Sun   Dinner    2
1    10.34    1.66   Male      No     Sun   Dinner    3
2    21.01    3.50   Male      No     Sun   Dinner    3
3    23.68    3.31   Male      No     Sun   Dinner    2
4    24.59    3.61   Female    No     Sun   Dinner    4
```

```
>>> tips.size.unique()
Traceback (most recent call last):
  File "<stdin>", line 1, in <module>
AttributeError: 'numpy.int32' object has no attribute 'unique'
>>> tips.size
1708
>>> len(tips) * len(tips.columns)
1708
>>> tips['size'].unique()
array([2, 3, 4, 1, 6, 5], dtype=int64)
>>> sns.relplot(x="total_bill", y="tip", hue="size", data=tips);
<seaborn.axisgrid.FacetGrid object at 0x00000244D49628F0>
```

从图17.33可以看到，聚餐人数(size)的颜色随着数值增加而变深，Seaborn将值转换为浮点数，并使用连续调色板(sequential palette)进行颜色处理。

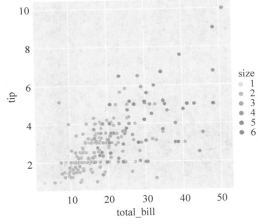

图17.33　使用Seaborn库中的relplot()方法绘制带有hue参数的聚餐人数的散点图

```
>>> sns.relplot(x="total_bill", y="tip", size="size", data=tips);
<seaborn.axisgrid.FacetGrid object at 0x00000244D4999090>
```

在演示了如何使用hue参数进行第三个变量设置之后，我们还可以使用size参数，示例代码如上。与前面的示例一样，我们将使用size变量作为size参数的输入，这将根据size这个第三变量改变点的大小，如图17.34所示。

我们可以使用sizes变量扩展前面的示例，该变量可以是列表、字典或元组的形式。对于分类变量来说，字典将包含相关值的级别，因此对于smoker变量，可以按如下方式传递字典数据：

```
>>> tips.smoker.unique()
['No', 'Yes']
Categories (2, object): ['Yes', 'No']
```

```
>>> sizes_dict = {'No': 100, 'Yes': 200}
>>> sns.relplot(x = "total_bill", y = "tip", size = "smoker",
... sizes = sizes_dict, data = tips)
<seaborn.axisgrid.FacetGrid object at 0x000001FB791B1840>
```

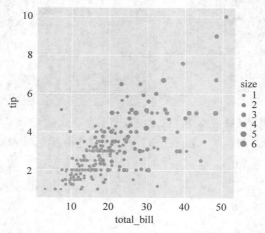

图 17.34　使用 Seaborn 库中的 relplot() 方法绘制带有 size 参数的聚餐人数散点图

我们也可以用列表代替字典，示例代码如下：

```
>>> sns.relplot(x = "total_bill", y = "tip", size = "smoker",
... sizes = [100,200], data = tips)
<seaborn.axisgrid.FacetGrid object at 0x000001FB7974DF90>
```

如果 size 用数值表示，我们可以给 sizes 参数传递一个包含最小值和最大值的元组，然后根据不同的数值显示每个点合适的大小，如图 17.35 所示。

```
>>> sns.relplot(x = "total_bill", y = "tip", size = "size",
... sizes = (15, 200), data = tips);
<seaborn.axisgrid.FacetGrid object at 0x000001FB791B1840>
```

接下来，我们将查看 Seaborn 中的折线图。最初，我们将使用 NumPy 中的两个方法设置 DataFrame。第一个是采用 arange() 方法创建从 $0 \sim n-1$ 的升序排列的整数列表，示例如下：

```
>>> np.arange(10)
array([0, 1, 2, 3, 4, 5, 6, 7, 8, 9])
```

另一种方式是采用 random 模块中的 randn() 方法生成一组随机数，并应用 cumsum() 方法对这些随机数进行累加。

```
>>> np.random.randn(10)
array([-0.45897172, -0.61704119, -0.25782623, 1.45683121, -0.18178387,
```

```
                1.25582391, 2.49150883, 0.77940559, -0.26918884, 0.03801411])
>>> np.random.randn(10).cumsum()
array([-0.08257024, -1.60249805, 0.18510565, -0.15252165, -0.24339787,
       -2.42919975, -3.02106779, -2.50762737, -2.58124245, -3.88439594])
```

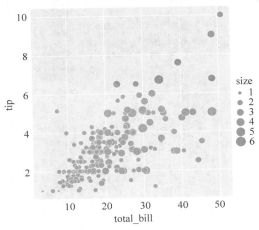

图 17.35 使用 Seaborn 库中的 relplot()方法绘制不同散点尺寸的散点图

注意，上述示例中的值不会相加，因为每次生成一组随机数时都会得到不同的值，而生成的随机数也没有存储。我们可以将第一组随机数赋值给一个变量，此时可以看到数据的累加效果，如下所示：

```
>>> x = np.random.randn(10)
>>> print(x)
[ 0.33964753 -1.30018827 1.47923716 -0.38759109 0.51351698 1.53193801
  2.38749607 -0.73040404 -1.12189689 1.23649563]
>>> x.cumsum()
array([ 0.33964753, -0.96054074, 0.51869642, 0.13110533, 0.64462231,
        2.17656032, 4.56405639, 3.83365235, 2.71175546, 3.94825109])
```

利用上述方法，我们可以创建一个 DataFrame，将一些随机数作为列值。为了将数据绘制出来，可以使用 lineplot()方法或使用 relplot()方法（将 kind 参数设置为"line"）。示例代码如下，绘制的图形如图 17.36 所示。

```
>>> import numpy as np
>>> import pandas as pd
>>> import matplotlib.pyplot as plt
>>> import seaborn as sns
>>> sns.set(style = "darkgrid")
>>> df = pd.DataFrame(dict(time = np.arange(500),
value = np.random.randn(500).cumsum()))
>>> df.head()
```

```
        time      value
0       0       - 0.960025
1       1       - 1.322551
2       2       - 1.108571
3       3       - 2.087655
4       4       - 2.752729
>>> g = sns.relplot(x = "time", y = "value", kind = "line", data = df)
>>> g.fig.autofmt_xdate()
```

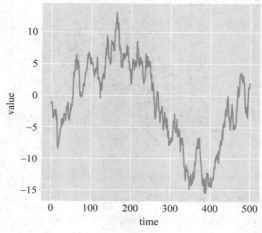

图 17.36　使用 Seaborn 库中的 relplot()方法绘制折线图

接下来,我们引入 fmri 数据集①,并讲解如何使用 relplot()方法绘制 signal 与 timepoint 之间的折线图。通过数据集可以看出每个 timepoint 都有多个测量值,示例代码如下②:

```
>>> fmri = sns.load_dataset("fmri")
>>> fmri[fmri['timepoint'] == 18].head()
   subject    timepoint    event    region      signal
0   s13       18           stim     parietal    - 0.017552
2   s12       18           stim     parietal    - 0.081033
3   s11       18           stim     parietal    - 0.046134
4   s10       18           stim     parietal    - 0.037970
5   s9        18           stim     parietal    - 0.103513
```

---

① fmri 是 Seaborn 自带的功能性磁共振成像数据集,包含 1064 行数据,共 5 列。每列数据含义如下:①subject 表示刺激类型,共有 14 种;②timepoint 表示时间点,共有 19 种;③event 表示事件,共 2 种;④region 表示大脑区域,共 2 种;⑤signal 表示信号,共 1064 个结果。

② 执行 sns.load_dataset("fmri")代码时可能会出现 URLError 或[Errno 11004]或 RemoteDisconnected 等错误,可能是由于网络问题导致访问 Seaborn 的数据集时无法下载,此时可以到网上下载 fmri.csv 数据集文件,Windows 系统可以复制到本地 seaborn-data 目录,如 C:\Users\用户名\seaborn-data(译者使用 Anaconda 或 Spyder)。

```
>>> fmri.groupby('timepoint').count()['signal']
>>> timepoint
0     56
1     56
2     56
3     56
4     56
5     56
6     56
7     56
8     56
9     56
10    56
11    56
12    56
13    56
14    56
15    56
16    56
17    56
18    56
Name: signal, dtype: int64
```

因此，对于每个 timepoint 都有 56 个值。现在，将这些值传递到 relplot() 函数中绘制折线图，结果如图 17.37 所示。图中的折线表示 timepoint 的平均值，阴影表示这些点 95% 的置信区间。

```
sns.relplot(x = "timepoint", y = "signal", kind = "line", data = fmri);
```

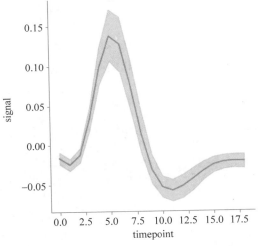

图 17.37　使用 relplot() 方法绘制带有平均值和置信区间的折线图

我们可以通过将 relplot() 函数中的 ci 参数设为 None 来消除置信区间,绘制结果如图 17.38 所示。

```
sns.relplot(x = "timepoint", y = "signal", ci = None,
kind = "line", data = fmri);
```

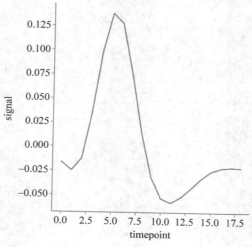

图 17.38　使用 relplot() 方法绘制有平均值但没有置信区间的折线图

通过将 relplot() 函数中的 ci 参数设置为 sd 可以改变平均值附近的阴影区间。因此,我们可以通过传递 sd 来改变平均值周围的间隔,重新计算平均值附近的标准偏差,绘制结果如图 17.39 所示。

```
sns.relplot(x = "timepoint", y = "signal", kind = "line",
ci = "sd", data = fmri);
```

就像前面在散点图中看到的那样,可以在 relplot() 函数中使用 hue 参数传递第三个变量来对数据进行分组。在下面的代码中,我们利用 event 变量将数据分成两组,绘制结果如图 17.40 所示。

```
>>> fmri.head()
   subject  timepoint  event  region    signal
0  s13      18         stim   parietal  -0.017552
1  s5       14         stim   parietal  -0.080883
2  s12      18         stim   parietal  -0.081033
3  s11      18         stim   parietal  -0.046134
4  s10      18         stim   parietal  -0.037970
>>> fmri.event.unique()
>>> sns.relplot(x = "timepoint", y = "signal", hue = "event",
kind = "line", data = fmri);
```

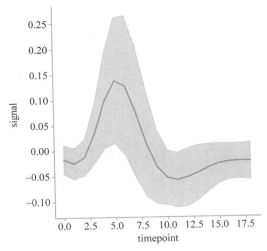

图 17.39 使用 relplot()方法绘制带有平均值和标准偏差的折线图

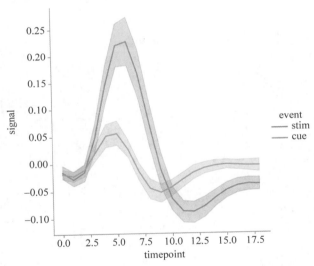

图 17.40 使用带有 hue 参数的 relplot()方法绘制折线图

我们再对上述示例进行扩展,在 relplot()函数中将 hue 参数设置为 region,将 style 参数设置为 event。此时,我们用不同的颜色代表每个 region,使用不同的线型代表不同的 event。这样就可以在折线图中表示更多的变量,如图 17.41 所示。

```
>>> fmri.region.unique()
array(['parietal', 'frontal'], dtype=object)
>>> sns.relplot(x="timepoint", y="signal", hue="region", style="event",
kind="line", data=fmri);
```

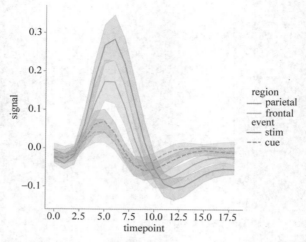

图 17.41 使用带有 hue 参数和 style 参数的 relplot()方法绘制折线图

下面采用另一个数据集来更改示例,此处将数据集修改为 dots 数据集[1],下述代码绘制结果如图 17.42 所示。

```
>>> dots = sns.load_dataset("dots").query("align == 'dots'")
>>> dots.head()
    align   choice   time    coherence    firing_rate
0   dots    T1       -80     0.0          33.189967
1   dots    T1       -80     3.2          31.691726
2   dots    T1       -80     6.4          34.279840
3   dots    T1       -80     12.8         32.631874
4   dots    T1       -80     25.6         35.060487
>>> dots.tail()
    align   choice   time    coherence    firing_rate
389 dots    T2       680     3.2          37.806267
390 dots    T2       700     0.0          43.464959
391 dots    T2       700     3.2          38.994559
392 dots    T2       720     0.0          41.987121
393 dots    T2       720     3.2          41.716057
>>> dots.coherence.unique()
>>> dots.choice.unique()
>>> sns.relplot(x = "time", y = "firing_rate",
hue = "coherence", style = "choice",
kind = "line", data = dots);
```

图 17.42 所示图形使用 time 和 firing_rate 作为 x 和 y 变量,然后将 hue 参数设置为 coherence,将 style 参数设置为 choice。同时使用 hue 和 style 可以让每个(x,y)点通过颜色和线型进行组合显示。因此,本质上来说,我们可以在一张图上显示 4 个变量。需

---

[1] dots 数据集也是 Seaborn 自带的数据集,共有 393 条数据,每条数据有 align、choice、time、coherence 和 firing_rate 等 5 个变量。

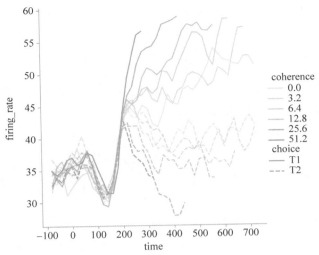

图 17.42　基于 dots 数据集使用带有 hue 参数和 style 参数的 relplot() 方法绘制折线图

要注意的是，dots 数据集特别适合这种图形的绘制，因为 coherence 变量有 6 种取值，choice 变量有 2 种取值，所以二者组合可以有 12 条线。当考虑采用这种类型的绘图时，需要确定数据集是否合适，绘图结果应该能对数据进行很好的展示（而不是展示的信息更少）。

到目前为止，我们已经介绍了单图的绘制方法。但是，如果我们想要在同一幅图上比较多个变量之间的关系，该怎么办呢？下面以 tips 数据集为例，来看一下消费总金额（total bill）小费金额（tip）之间的关系。

```
>>> import numpy as np
>>> import pandas as pd
>>> import matplotlib.pyplot as plt
>>> import seaborn as sns
>>> sns.set(style = "darkgrid")
>>> tips = sns.load_dataset("tips")
>>> tips.head()
   total_bill   tip    sex     smoker  day   time    size
0  16.99        1.01   Female  No      Sun   Dinner  2
1  10.34        1.66   Male    No      Sun   Dinner  3
2  21.01        3.50   Male    No      Sun   Dinner  3
3  23.68        3.31   Male    No      Sun   Dinner  2
4  24.59        3.61   Female  No      Sun   Dinner  4
>>> tips.time.unique()
['Dinner', 'Lunch']
Categories (2, object): ['Lunch', 'Dinner']
>>> tips.smoker.unique()
['No', 'Yes']
Categories (2, object): ['Yes', 'No']
>>> sns.relplot(x = "total_bill", y = "tip", hue = "smoker",
      col = "time", data = tips);
```

上述示例中,我们已经研究了消费总金额和小费金额之间的关系,这已经在前面讨论过了。下面看一下数据集中的其他变量,我们可以使用时间段(time)和吸烟顾客(smoker)进一步研究消费总金额和小费金额之间的关系。将 hue 参数设置为 smoker 可以将吸烟顾客的数据区分开来,另外,借助 col 参数可以进一步扩展绘图功能。在本例中,我们将 col 参数设置为 time,与 smoker 一样,time 也只包含两个类别。通过并排显示有助于展示变量之间的关系,如图 17.43 所示。

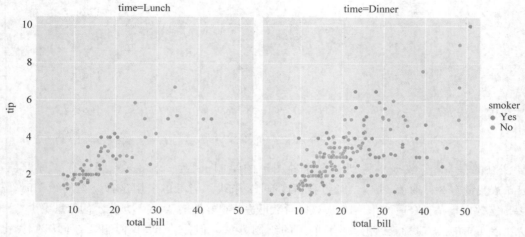

图 17.43　基于 tips 数据集绘制多散点图

通过将行和列生成多重绘图可以对前面示例进行扩展。在此,继续从 Seaborn 加载 fmri 数据集。

```
>>> fmri = sns.load_dataset("fmri")
>>> fmri.head()
  subject    timepoint    event    region     signal
0 s13        18           stim     parietal   -0.017552
1 s5         14           stim     parietal   -0.080883
2 s12        18           stim     parietal   -0.081033
3 s11        18           stim     parietal   -0.046134
4 s10        18           stim     parietal   -0.037970
>>> fmri.dtypes
subject        object
timepoint      int64
event          object
region         object
signal         float64
dtype: object
>>> fmri.event.unique()
array(['stim', 'cue'], dtype=object)
>>> fmri.region.unique()
array(['parietal', 'frontal'], dtype=object)
>>> fmri.subject.unique()
```

```
array(['s13', 's5', 's12', 's11', 's10', 's9', 's8', 's7', 's6', 's4',
       's3', 's2', 's1', 's0'], dtype=object)
>>> sns.relplot(x = "timepoint", y = "signal", hue = "subject",
col = "region", row = "event", height = 3,
kind = "line", estimator = None, data = fmri);
```

与前面一样，在上述代码中首先使用 head() 函数检查数据集。本例中进一步使用 dtypes 方法显示每列的数据类型，可以看出，timepoint 和 signal 的数据类型分别为 int64 和 float64，因此这两个变量是作为 x 和 y 变量的理想选择。观察其他变量，我们发现 event 和 region 只有 2 个不同的值，而 subject 有 14 个不同的数值。因此，将这些变量使用同一幅图进行绘制时，最适合利用 hue 参数区分变量颜色，并使用 event 和 region 作为行和列变量。在代码中，将参数 col 设置为 region，将参数 row 设置为 event。最终得到了一幅 2×2 的图，其小标题包含了不同的变量组合，如图 17.44 所示。

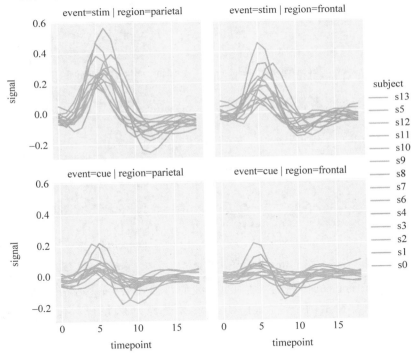

图 17.44 基于 fmri 数据集绘制带有行和列标题的多折线图

这幅图绘制得很好，但是由于 hue 参数有 14 个变量所以很难从颜色上区分。这一问题也好解决，我们可以通过选择数据子集来减少这种情况，如图 17.45 所示。

```
>>> fmri['subject'].isin(['s0','s1','s2']).head()
Out[40]:
0    False
1    False
```

```
2    False
3    False
4    False
Name: subject, dtype: bool
>>> fmri_red = fmri[fmri['subject'].isin(['s0','s1','s2'])]
>>> sns.relplot(x = "timepoint", y = "signal", hue = "subject",
col = "region", row = "event", height = 3,
kind = "line", estimator = None, data = fmri_red);
```

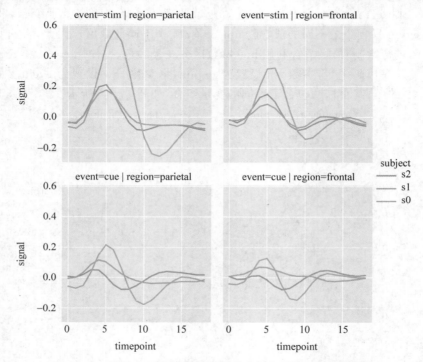

图 17.45　基于 fmri 数据子集绘制带有行和列标题的多折线图

如果我们只想通过 col 传递的一个变量，我们可以设置 col_wrap 参数，该参数给出并排绘图的最大值。在本例中，将 col_wrap 参数设置为 5，每行将包含 5 个子图，共 3 行，以适合 14 个 subject 取值。这为我们展示了仅使用单个变量设置行和列的效果，如图 17.46 所示。

```
sns.relplot(x = "timepoint", y = "signal", hue = "event", style = "event",
col = "subject", col_wrap = 5,
height = 3, aspect = .75, linewidth = 2.5,
kind = "line", data = fmri.query("region == 'frontal'"));
```

接下来，我们考虑绘制分类数据（categorical data），首先借助 tips 数据集介绍绘制分类统计图方法 catplot() 的使用，示例代码如下：

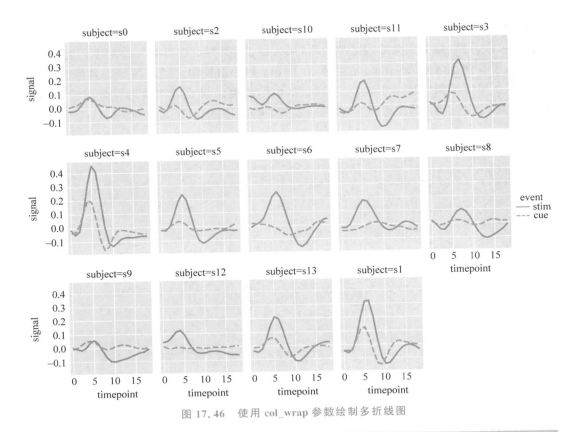

图 17.46 使用 col_wrap 参数绘制多折线图

```
>>> tips = sns.load_dataset("tips")
>>> tips.head()
   total_bill   tip    sex     smoker  day   time    size
0  16.99        1.01   Female  No      Sun   Dinner  2
1  10.34        1.66   Male    No      Sun   Dinner  3
2  21.01        3.50   Male    No      Sun   Dinner  3
3  23.68        3.31   Male    No      Sun   Dinner  2
4  24.59        3.61   Female  No      Sun   Dinner  4
>>> tips.dtypes
total_bill   float64
tip          float64
sex          category
smoker       category
day          category
time         category
size         int64
dtype: object
>>> tips['day'].unique()
['Sun', 'Sat', 'Thur', 'Fri']
Categories (4, object): ['Thur', 'Fri', 'Sat', 'Sun']
>>> sns.catplot(x = "day", y = "total_bill", data = tips);
```

上述代码加载了 Seaborn 自带的 tips 数据集,并使用 head()和 dtypes()方法查看其内容。可以看到,分类数据的数据类型为 category,该类型数据具有一些选项。对于 day 列数据,需要注意该列为 category 类型,并且 day 列的取值包含了一些星期名称的缩写。然后,我们可以绘制按星期几划分的消费总金额,绘制结果是以按星期几分组的点的形式,并用颜色对分组结果进行区分,如图 17.47 所示。

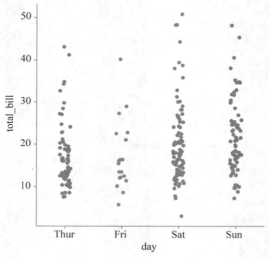

图 17.47 使用 catplot()方法绘制 tips 数据集中消费总金额与星期几之间的关系图

现在,如果考虑这些数据,则会看到一些重复数据。

```
>>> tips[tips['day'] == 'Thur'].sort_values(by = 'total_bill').head(20)
     total_bill  tip    sex     smoker  day    time    size
149  7.51        2.00   Male    No      Thur   Lunch   2
195  7.56        1.44   Male    No      Thur   Lunch   2
145  8.35        1.50   Female  No      Thur   Lunch   2
135  8.51        1.25   Female  No      Thur   Lunch   2
126  8.52        1.48   Male    No      Thur   Lunch   2
148  9.78        1.73   Male    No      Thur   Lunch   2
82   10.07       1.83   Female  No      Thur   Lunch   1
136  10.33       2.00   Female  No      Thur   Lunch   2
196  10.34       2.00   Male    Yes     Thur   Lunch   2
117  10.65       1.50   Female  No      Thur   Lunch   2
132  11.17       1.50   Female  No      Thur   Lunch   2
128  11.38       2.00   Female  No      Thur   Lunch   2
120  11.69       2.31   Male    No      Thur   Lunch   2
147  11.87       1.63   Female  No      Thur   Lunch   2
133  12.26       2.00   Female  No      Thur   Lunch   2
118  12.43       1.80   Female  No      Thur   Lunch   2
124  12.48       2.52   Female  No      Thur   Lunch   2
201  12.74       2.01   Female  Yes     Thur   Lunch   2
202  13.00       2.00   Female  Yes     Thur   Lunch   2
198  13.00       2.00   Female  Yes     Thur   Lunch   2
```

由上可见,周四消费总金额 13.00 出现了重复数据。在前面绘制的散点图中没有考虑到这一点,因此需要采取某种方法来处理这一问题。幸运的是,我们有一个设置可以用来改变这一点。通过将 kind 参数设置为 swarm,可以在算法上防止变量的重叠,这为我们提供了更好的数据分布表示,如图 17.48 所示。应该注意的是,kind 参数的默认值是 strip,也就是在上一示例中所绘制的图形样式。

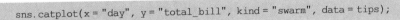

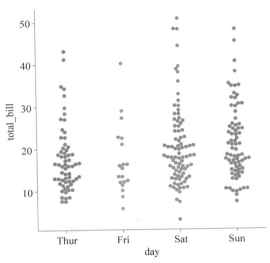

图 17.48　使用 catplot() 方法绘制 tips 数据集中消费总金额与聚餐人数之间的关系,其中 kind 参数设为 swarm

前面我们已经介绍了怎样根据数值进行分组图形绘制,并使用色调对分组进行区分,对于 catplot() 绘图也同样适用。由变量列表可以看到 sex 是另一个分类变量,将改变量传递给 hue 参数可以很好地进行数据展示。本例所绘制的图形与之前一张绘图的区别之处在于颜色现在由 hue 参数驱动,而其他分类变量通过 x 参数进行传递,如图 17.49 所示。

```
>>> tips.dtypes
>>> tips['sex'].unique()
>>> sns.catplot(x = "day", y = "total_bill", hue = "sex",
kind = "swarm", data = tips);
```

到目前为止,我们已经处理了文本数据形式的分类变量,如 sex 或 date,但如果类别是数字形式该怎样处理呢？在下一个示例中,我们将 x 值设置为 size(聚餐人数),这是一个整数,从示例可以看到,该值被作为分类数据进行处理,与之前的示例类似。在代码中,我们引入了一个新的 DataFrame 方法,即 query() 方法,该方法对 DataFrame 执行一个字符串形式的查询。本例中,输入的查询条件为"size != 3",这意味着希望得到聚餐人数不等于 3 的数据。这也可以使用布尔序列来实现,此时可以将 data 参数设置为"tips

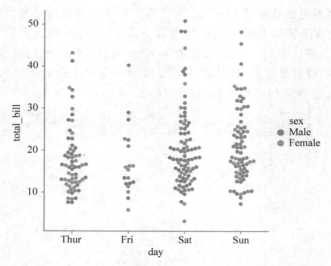

图 17.49　使用 catplot() 方法绘制 tips 数据集中消费总金额与聚餐人数之间的关系，其中 kind 参数设为 swarm，hue 参数设为 sex

[tips['size']=3]"。需要注意的是，catplot() 绘图将根据聚餐人数的大小对 x 值进行适当排序。

```
>>> sns.catplot(x = "size", y = "total_bill", kind = "swarm",
    data = tips.query("size != 3"));
```

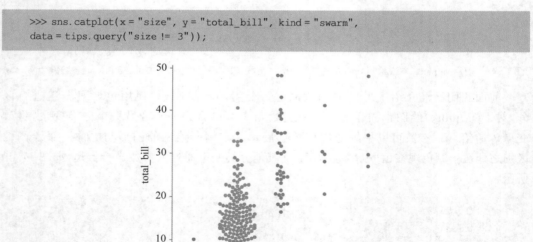

图 17.50　使用 catplot() 方法绘制消费总金额与聚餐人数之间的关系

　　数值排序问题相对简单，因为借助 Seaborn 可以使用升序来排序数值数据，但是对于分类变量，并不能直接进行排序。如果我们要绘制吸烟顾客与小费金额的关系，那么如何控制吸烟顾客的顺序[即吸烟(Yes)和不吸烟(No)的顺序]呢？为此，可以使用 order

参数,并将其设置为期望的显示顺序,本例中 order 赋值为"No"和"Yes"组成的列表,绘图结果如图 17.51 所示。

```
>>> tips.smoker.unique()
['No', 'Yes']
Categories (2, object): ['Yes', 'No']
>>> sns.catplot(x = "smoker", y = "tip", order = ["No", "Yes"],
data = tips);
```

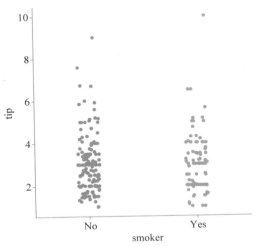

图 17.51　catplot()中使用 order 参数绘制吸烟顾客与小费金额的关系

下面进一步对示例进行修改,将 x 轴修改为 total_bill,将 y 轴设置为分类变量 day,将 kind 参数设置为 swarm,此时,绘图样式从垂直转化为水平绘图。在本例中,我们还应用了 hue 参数和 swarm 类型,这表明可以使用水平或垂直绘制同样结果的图形,如图 17.52 所示。

```
>>> sns.catplot(x = "total_bill", y = "day", hue = "time",
kind = "swarm", data = tips);
```

前面的示例都绘制了散点图,然而,通过改变传递给 kind 参数的数据可以生成不同类型的图形。首先来看一下如何将已有数据绘制成箱线图。为此,我们将 kind 参数设置为"box",它将把默认的散点图转换为箱线图。在下面的示例中,x 轴设置为 day,y 轴设置为 total_bill,示例代码如下:

```
>>> sns.catplot(x = "day", y = "total_bill", kind = "box", data = tips);
```

这样做的结果会得到一个标准的箱线图,如图 17.53 所示。但图表之外的数据显示为数据点,每个数据点都显示在图表上面。

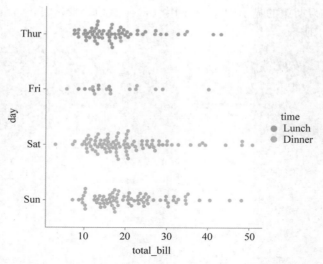

图 17.52　catplot()中使用 hue 和 kind 参数绘制总消费金额与星期几之间的关系

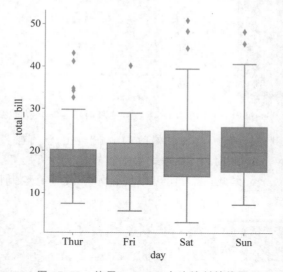

图 17.53　使用 catplot()方法绘制箱线图

正如前面的散点图示例所示，我们可以添加 hue 参数，使用颜色为每个类别提供不同的箱线图，用色调类区分数据层级，如图 17.54 所示。

```
>>> sns.catplot(x = "day", y = "total_bill", hue = "smoker",
kind = "box", data = tips);
```

下面介绍的绘图类型是增强型箱线图(boxen plot)，此时需要将 kind 参数设置为

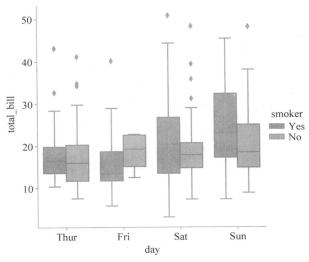

图 17.54　使用带有 hue 参数的 catplot( )绘制箱线图

boxen。下面的示例使用了 diamonds 数据集①，绘制了钻石价格（price）与钻石颜色（color）之间的关系图形。该图与箱线图类似，不同之处在于数据分组超出了箱线图所示的四分位数，如图 17.55 所示。这样可以更好地了解数据的分布情况，给出更多的数据分组。

```
>>> diamonds = sns.load_dataset("diamonds")
>>> diamonds.head()
   carat      cut color clarity  depth  table  price     x     y     z
0   0.23    Ideal     E     SI2   61.5   55.0    326  3.95  3.98  2.43
1   0.21  Premium     E     SI1   59.8   61.0    326  3.89  3.84  2.31
2   0.23     Good     E     VS1   56.9   65.0    327  4.05  4.07  2.31
3   0.29  Premium     I     VS2   62.4   58.0    334  4.20  4.23  2.63
4   0.31     Good     J     SI2   63.3   58.0    335  4.34  4.35  2.75
>>> diamonds.dtypes
carat      float64
cut       category
color     category
clarity   category
depth      float64
table      float64
price        int64
x          float64
```

---

①　diamonds 是 Seaborn 自带的数据集，共有 10 个变量，53940 行数据。每行数据代表一个钻石的属性，一般用于研究钻石价格与重量、形状、切割状态、颜色、透明度之间的关系。其中 10 个变量含义如下：carat 为钻石的重量；cut 为钻石的切工，由低到高依次为 Fair、Good、Very Good、Premium、Ideal；color 为钻石颜色，取值从最低的 J 到最高的 D；clarity 是钻石的纯净度，从低到高依次为 I1、SI1、SI2、VS1、VS2、VVS1、VVS2、IF；depth 为深度比例；table 是钻石的台面比例；price 为钻石的价格；x、y、z 分别为钻石的长、宽、高。

```
y          float64
z          float64
dtype: object
>>> diamonds.color.unique()
['E', 'I', 'J', 'H', 'F', 'G', 'D']
Categories (7, object): ['D', 'E', 'F', 'G', 'H', 'I', 'J']
>>> sns.catplot(x = "color", y = "price", kind = "boxen",
    data = diamonds.sort_values("color"));
```

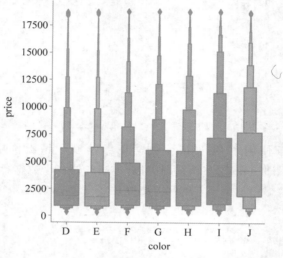

图 17.55　使用 catplot() 绘制箱线图

上述示例给出的增强箱线图比一般箱线图能更好地展示多个数据分组时的分布情况。在下一个示例中，我们可以在获得箱线图概要的同时显示箱线图的信息。为此，我们将 kind 参数设置为"violin"来绘制小提琴图（violin plot）。示例中，我们与以前一样绘制消费总金额（total bill）与星期几（day）之间的关系，将 hue 参数设置为时间段变量"time"，该变量具有晚餐（Dinner）和午餐（Lunch）两个唯一值。图中绘制了相关数据的箱线图，并计算了核密度估计值（KDE）。借助 KDE，我们将能获取数据的更多细节。与直方图相似，KDE 展示数据的分布情况。对于直方图，我们需要设置 bin 数量以确定绘图的外观，但由于小提琴绘图的 KDE 无法设置，所以需要设置平滑参数。在下面给出的示例中，可以看到平滑参数（bw_adjust）设置为 0.15，其默认值为 1。选择该参数是确定绘图外观的关键，因为过度平滑可能会去除数据集的某些方面的信息。本例中使用的另外一个参数是 cut 参数（该参数以前没有用过，用法很简单），它决定了与平滑参数一起使用时绘图将如何扩展。在本例中，cut 参数设置为 0，这意味着要将小提琴图范围限制在观察数据的范围内。

在该示例中，我们可以看到不是每一个星期几（day）和聚餐时间（time）都有数据点，因此，周四和周五有两个小提琴图，而周六和周日只有一把小提琴图。另一个有趣的结果是单数据点的处理,因为周四晚餐只有一个数据点。在这种情况下,total_bill 为 18.78 点处有一条直线,代表消费总金额。如果不希望出现这一直线,可以在绘制之前从数据中删除周四晚餐的这一数据点。最终绘制结果如图 17.56 所示。

```
>>> tips.head()
   total_bill   tip    sex     smoker  day   time     size
0  16.99        1.01   Female  No      Sun   Dinner   2
1  10.34        1.66   Male    No      Sun   Dinner   3
2  21.01        3.50   Male    No      Sun   Dinner   3
3  23.68        3.31   Male    No      Sun   Dinner   2
4  24.59        3.61   Female  No      Sun   Dinner   4
>>> tips.time.unique()
['Dinner', 'Lunch']
Categories (2, object): ['Lunch', 'Dinner']
>>> tips.groupby(by = ['day', 'time']).count()
            total_bill  tip  sex  smoker  size
day  time
Thur Lunch  61          61   61   61      61
     Dinner 1           1    1    1       1
Fri  Lunch  7           7    7    7       7
     Dinner 12          12   12   12      12
Sat  Lunch  0           0    0    0       0
     Dinner 87          87   87   87      87
Sun  Lunch  0           0    0    0       0
     Dinner 76          76   76   76      76
>>> thursday = tips[(tips['day'] == 'Thur')]
>>> thursday.head()
    total_bill  tip    sex    smoker  day   time    size
77  27.20       4.00   Male   No      Thur  Lunch   4
78  22.76       3.00   Male   No      Thur  Lunch   2
79  17.29       2.71   Male   No      Thur  Lunch   2
80  19.44       3.00   Male   Yes     Thur  Lunch   2
81  16.66       3.40   Male   No      Thur  Lunch   2
>>> thursday[thursday['time'] == 'Dinner']
     total_bill  tip   sex     smoker  day   time    size
243  18.78       3.0   Female  No      Thur  Dinner  2
>>> sns.catplot(x = "total_bill", y = "day", hue = "time",
kind = "violin", bw_adjust = 0.15, cut = 0,
data = tips);
```

如果将 split 参数设置为 True,并应用 hue 参数,我们会看到在同一个小提琴图中显示两组值,这是一个有趣的绘图变化。在下面的示例中,每一个星期几(day)都有一幅小提琴图,而不是通常使用 hue 参数时预期的两幅图。此时,我们是在整个数据上绘制箱线图,并在小提琴图使用 split 参数,这样便可以将数据总体分布与每个数据类别的分布进行色调上的比较。该示例将平滑参数 bw_adjust 设置为默认值,可以与上一示例中应用 bw_adjust 参数的效果进行比较。绘图结果如图 17.57 所示。

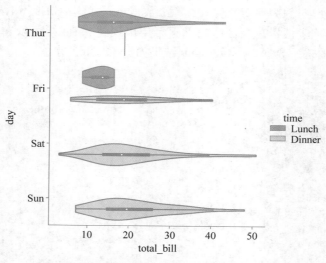

图 17.56 使用 catplot() 绘制小提琴图

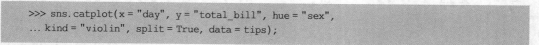

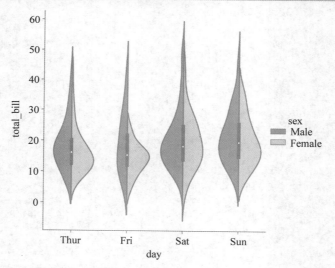

图 17.57 使用 catplot() 绘制色调分割的小提琴图

接下来将研究如何使用条形图捕捉数据集中分类数据的变化。为了实现这一点，需要将参数 king 设置为"bar"，这样就会基于 x 和 y 变量生成一幅条形图，其中 x 变量设置为"sex"，y 变量设置为"survived"；并将参数 hue 设置为"class"，绘制的条形图如图 17.58 所示。条形图默认显示数据的平均值，如果有多个观测值，则需从数据中提取置信区间，并在条形图顶部显示为垂直线。

```
>>> import seaborn as sns
>>> import matplotlib.pyplot as plt
>>> sns.set(style = "ticks", color_codes = True)
>>> tips = sns.load_dataset("tips")
>>> titanic = sns.load_dataset("titanic")
>>> titanic.head()
   survived  pclass     sex   age  ...  deck  embark_town  alive  alone
0         0       3    male  22.0  ...   NaN  Southampton     no  False
1         1       1  female  38.0  ...     C    Cherbourg    yes  False
2         1       3  female  26.0  ...   NaN  Southampton    yes   True
3         1       1  female  35.0  ...     C  Southampton    yes  False
4         0       3    male  35.0  ...   NaN  Southampton     no   True
[5 rows x 15 columns]
>>> titanic.dtypes
Out[28]:
survived         int64
pclass           int64
sex             object
age            float64
sibsp            int64
parch            int64
fare           float64
embarked        object
class         category
who             object
adult_male        bool
deck          category
embark_town     object
alive           object
alone             bool
dtype: object
>>> sns.catplot(x = "sex", y = "survived", hue = "class", kind = "bar", data = titanic);
```

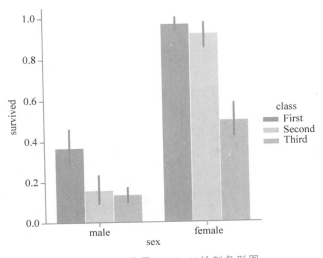

图 17.58　使用 catplot( ) 绘制条形图

如果是对频率而不是对平均值感兴趣,可以将 kind 参数设置为"count",该参数可以绘制如图 17.59 所示的计数图(count plot)。在此,我们没有传递 x 值,而是使用 hue 参数对变量 class 进行分组,并将 y 值设置为"deck"。此时,变量 deck 按 class 进行分组,每组的数量就是 x 轴的坐标。

```
>>> sns.catplot(y = "deck", hue = "class", kind = "count",
    palette = "pastel", edgecolor = ".6",
    data = titanic);
```

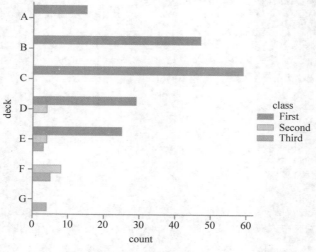

图 17.59 使用 catplot()绘制计数图

到目前为止,所给示例都是研究显示传递给绘图的变量之间的关系。以 iris 数据集为例,我们可以更加轻松地将数据传递给某一方法,以便在整个数据集中应用。在下面的示例中,数据共有 5 列,其中 4 列的数据类型为 float,将其传递到 catplot()中,并将 kind 参数设置为"box",从而生成这 4 个 float 类型变量的箱线图。因此,我们更加灵活地使用 Seaborn 进行绘图,如图 17.60 所示。还应该注意,我们通过设置 orient 参数为"h"实现水平箱线图的绘制。

```
>>> iris = sns.load_dataset("iris")
>>> iris.head()
Out[37]:
   sepal_length  sepal_width  petal_length  petal_width  species
0  5.1           3.5          1.4           0.2          setosa
1  4.9           3.0          1.4           0.2          setosa
2  4.7           3.2          1.3           0.2          setosa
3  4.6           3.1          1.5           0.2          setosa
4  5.0           3.6          1.4           0.2          setosa
```

```
>>> iris.dtypes
Out[38]:
sepal_length    float64
sepal_width     float64
petal_length    float64
petal_width     float64
species         object
dtype: object
>>> sns.catplot(data = iris, orient = "h", kind = "box");
```

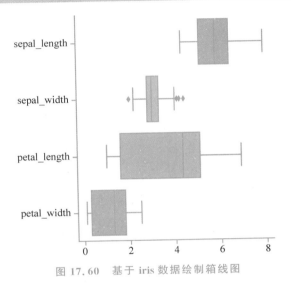

图 17.60　基于 iris 数据绘制箱线图

前面讲解了如何在 catplot() 上应用 hue 参数，下面介绍如何在 catplot() 上应用 col 参数。可以使用 col 参数创建多个子图。示例中依然使用 iris 数据集进行演示，并绘制了一幅星期几与消费总金额之间关系的蜂群图（swarm plot），其中 hue 参数设置为"smoker"，但为了扩展显示，我们添加了 col 参数，并将其设置为"time"，这样便得到了两幅子图，一幅的聚餐时间为午餐，另一幅的聚餐时间为晚餐，如图 17.61 所示。

```
>>> sns.catplot(x = "day", y = "total_bill", hue = "smoker",
    col = "time", aspect = .6,
    kind = "swarm", data = tips);
```

接下来将研究绘制一组数据而不是两个值之间的关系。这种图形称为单变量分布（univariate distribution），对于这类数据的显示，可能需要用到直方图（histogram）。为了说明这一点，可以简单地生成一些随机数据，然后将其传递到 distplot() 方法中，最后得到一幅适合数据的带有 KDE 的直方图，如图 17.62 所示。

```
>>> import numpy as np
>>> import pandas as pd
>>> import seaborn as sns
>>> import matplotlib.pyplot as plt
>>> from scipy import stats
>>> sns.set(color_codes = True)
>>> x = np.random.normal(size = 100)
>>> sns.distplot(x);
```

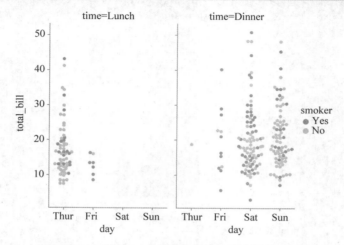

图 17.61　带有 col 参数的 catplot() 绘制多个子图

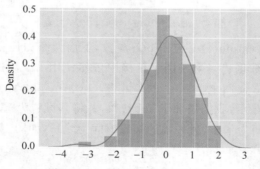

图 17.62　带有 KDE 的直方图

我们可以自定义直方图,可以通过将 kde 选项设置为 False 来取消核密度估计,并通过将 rug 选项设置为 True 来添加 rugplot 分布,如图 17.63 所示。rugplot 分布将每个数据值显示为沿 x 轴分布的小短线,从而为我们提供数据的表示。

```
sns.distplot(x, kde = False, rug = True);
```

前面给出的直方图示例中,bins 值采用了 distplot() 在数据集上的默认值。如果想指定要使用的箱的数量,只需将 bins 参数设置为所需的值,绘图将划分为指定的箱的数

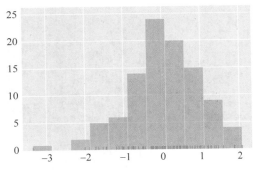

图 17.63　带有 rugplot 分布的直方图

量,如图 17.64 所示。

```
>>> x = np.random.normal(size = 100)
>>> x[0:10]
Out[27]:
array([ 0.59134269, -1.14907722, -1.21189156, 0.58493258, 0.99181633,
       -0.49552272, -0.59764092, -1.90194934, -0.18841865, -0.2439455 ])
>>> sns.distplot(x, bins = 20, kde = False, rug = True);
```

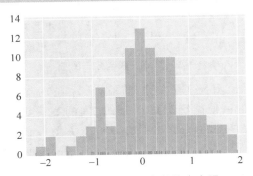

图 17.64　带有 bins 参数的直方图

前面已经介绍了散点图和直方图,现在来看看如何在一幅图上绘制这两种图。我们可以使用联合分布图(joint plot)方法,其中包含 x 对 y 的散点图以及 x 轴和 y 轴上每个变量的直方图。在下述示例中,首先创建了两个随机变量,并将它们放入 DataFrame 中。然后将此 DataFrame 作为 data 参数传入,并设置 x 和 y 参数为 DataFrame 中的 "x" 和 "y" 列,绘制的联合分布图如图 17.65 所示。

```
>>> x = np.random.normal(size = 100)
>>> y = np.random.normal(size = 100)
>>> df = pd.DataFrame({'x':x, 'y': y})
>>> df.head()
```

```
         x            y
0    0.323772    -0.521242
1    0.346981    -0.046482
2    0.745992    -0.257071
3    1.002763    -0.055071
4    0.456864    -0.943609
>>> sns.jointplot(x = "x", y = "y", data = df);
```

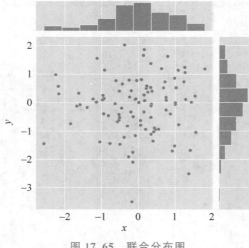

图 17.65　联合分布图

本章考虑的最后一种绘图是变量关系组图(pair plot)。在下述示例中，我们只需传入 iris 数据集的 DataFrame，便能绘制两两变量之间的关系图，如图 17.66 所示。其中，对角线上的图像是单变量的分布(显示的是变量的直方图)，而非对角线上的图像是两个变量之间的双变量散点图。需要注意的是，在示例中，iris 数据集的 DataFrame 中有分类数据，但利用 pair plot() 绘图时忽略了该列。

```
>>> iris = sns.load_dataset("iris")
>>> iris.head()
    sepal_length  sepal_width  petal_length  petal_width  species
0      5.1           3.5          1.4           0.2       setosa
1      4.9           3.0          1.4           0.2       setosa
2      4.7           3.2          1.3           0.2       setosa
3      4.6           3.1          1.5           0.2       setosa
4      5.0           3.6          1.4           0.2       setosa
>>> sns.pairplot(iris);
```

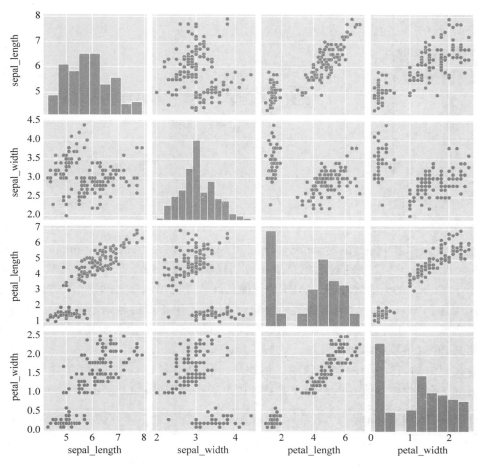

图 17.66　基于 iris 数据集的变量关系组图

本章小结

本章主要介绍了如何利用 Python 进行绘图，既讲解了简单绘图，也学习了非常复杂的可定制绘图。本书希望展示出 Python 在绘图方面的强大能力，书中示例可作为绘图的参考，并可作为提高 Python 技能的有用文档。

第 18 章 Python API

本章将介绍如何使用 Python 处理应用程序编程接口（application programming interfaces，API）。本章将通过示例介绍如何创建和访问 API。在开始编写代码之前，首先需要了解 API 的概念及用途。API 是一种允许软件应用程序之间通信的机制，可以实现应用程序和 Web 用户之间的通信。当前，API 的使用正变得越来越流行，它允许用户访问数据或与服务器通信。Python API 提供了一种统一的方法来实现这一点，从而被人们所熟悉，并进一步理解通信机理。

首先，创建自己的 API 进行说明。创建 API 需要用到 Python 中的 flask 包和 flask-restful 包，可以尝试导入这两个包查看系统中是否已经安装，如下所示：

```
>>> from flask import *
>>> from flask_restful import *
```

默认情况下，flask 包由 Python Anaconda 发行版提供，但 flask-restful 包可能会没有，如果是这种情况，则需要单独进行安装。可以在 flask-restful 包的官网进行下载安装，如图 18.1 所示。

本书编写的最初想法是不依赖于 Python Anaconda 发行版之外的内容就可以独立运行，但围绕 Anaconda 网站及相关链接的 URL（uniform resource locator，统一资源定位符）地址可能会发生变化，如果出现这种情况则只需要搜索 Anaconda 的包列表就可以得到，或者使用搜索引擎对 conda flask restful 进行简单搜索就可以找到相关的网页，然后就可以使用给定的 conda 命令从命令行安装 flask-restful 包。

所有包安装之后，我们可以创建第一个 API。下面将采用脚本形式创建一个名为 my_flask_api.py 的文件，并在文件中添加以下内容：

```
from flask import Flask
from flask_restful import Resource, Api

app = Flask(__name__)
```

```
api = Api(app)

class HelloWorld(Resource):
    def get(self):
        return {'hello': 'world'}

api.add_resource(HelloWorld, '/')

if __name__ == '__main__':
    app.run(debug = True)
```

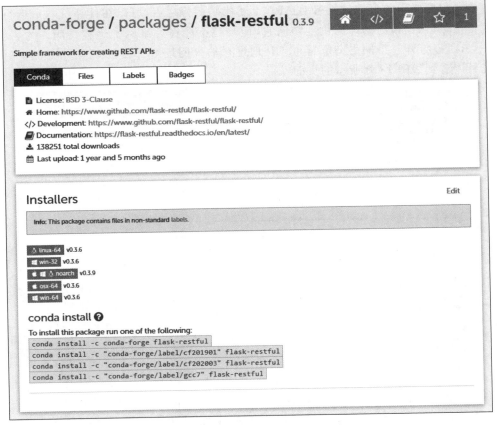

图 18.1　flask-restful 包下载页面示例

下面演示如何运行该段代码。首先,打开一个终端或命令提示符,并将目录更改为文件所在的位置。在命令行运行命令 Python my_flask_api.py 之后,得到如图 18.2 所示的内容[①]。

---

① 译者注:为验证代码的有效性,本章代码在 Windows 系统采用命令提示符或 PyCharm 进行验证,浏览器采用 Firefox 浏览器。

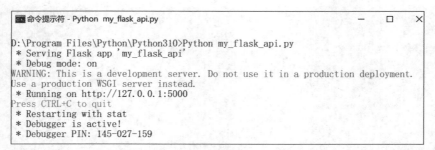

图 18.2 启动 API 时显示的终端界面

该段程序所做的事情就是启动 API，上述代码运行之后，可以在计算机本地运行它。也就是说，可以在自己的计算机上访问它，但是还不能在万维网上远程访问。为了演示这一点，可以打开一个网络浏览器，在 IP 地址栏输入网址 http://127.0.0.1:5000/，随后就可以看到如图 18.3 所示的页面。

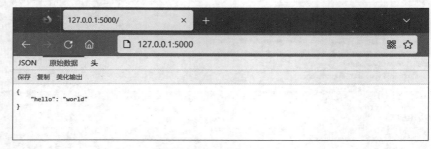

图 18.3 在浏览器显示 API

这是怎样实现的呢？让我们回过头来看看原始代码。

```
from flask import Flask
from flask_restful import Resource, Api

app = Flask(__name__)
api = Api(app)
```

上述代码中，首先从 flask 包和 flask-restful 包导入相关对象，然后使用 Flask() 方法创建了一个应用程序，当前模块的名称通过 __name__ 传入。这样，我们便创建了一个 app 对象，可以将该 app 对象传递给 Api() 方法，从而创建一个 api 对象。这一阶段属于设置阶段，接下来我们要向其中添加一个端点（endpoint）。

```
class HelloWorld(Resource):
    def get(self):
        return {'hello': 'world'}

api.add_resource(HelloWorld, '/')
```

在上述代码片段中，我们创建了一个名为 HelloWorld 的类，并带有一个名为

Resource 的参数。在该类中定义了一个名为 get 的方法，用于简单地返回一个字典数据：{'hello': 'world'}。

```
if __name__ == '__main__':
    app.run(debug = True)
```

最后给出的是启动 api 所执行的代码。对于 app 对象，我们使用其 run() 方法，并将 debug 参数设置为 True。这样，用户就可以在本地使用 api。这段代码包含了一行有趣的代码，即

```
if __name__ == '__main__':
```

该语句在 Python 中十分常见，但许多人不太理解为什么这样用，下面我们解释这一语句的用法。if 语句中使用"=="运算符判断两个变量的关系很容易理解，但 __name__ 和 '__main__' 是什么意思？如前所示，当利用 Flask() 创建对象时，__name__ 变量为我们提供当前模块的名称。但这是如何工作的呢？下面将通过创建两个文件的示例来展示，每个文件都调用 __name__ 变量，以显示其行为。首先创建第一个文件 file_one.py，代码如下：

```
out_str = 'file one __name__ is {}'.format(__name__)
print(out_str)
```

上述代码运行后，则会看到以下输出：

```
file one __name__ is __main__
```

现在，创建第二个文件 file_two.py，并输入以下代码：

```
out_str = 'file two __name__ is {}'.format(__name__)
print(out_str)
```

运行该代码后，则会看到以下输出：

```
file two __name__ is __main__
```

所有这些看起来都很合理，但如果将 file_two.py 导入 file_one.py，如下所示：

```
from file_two import *
out_str = 'file one __name__ is {}'.format(__name__)
print(out_str)
```

再次运行 file_one.py，得到以下输出：

```
file two __name__ is file_two
file one __name__ is __main__
```

这对于原始代码片段意味着什么呢？

```
if __name__ == '__main__':
```

从本质上讲，如果从该语句所在的脚本文件运行，那么__name__就会被赋值'__main__'，使if语句条件为真，执行if语句中的内容。

如果保持脚本my_flask_api.py一直运行，则可以在网址http://127.0.0.1:5000/运行API。现在，如果想以编程方式从自己的API获取数据，则需要使用requests包来访问数据。如果使用控制台(console)运行my_flask_api.py脚本文件，则可以使用以下代码进行交互方式访问在本地机器上运行的API[①]：

```
>>> import requests
>>> data = requests.get('http://127.0.0.1:5000/')
>>> data
<Response [200]>
>>> type(data)
<class 'requests.models.Response'>
```

现在，如果返回到运行API代码的窗口，可以看到，当运行以URL作为参数的requests.get()方法时向端点发出了get请求。需要注意的是，由requests.get()方法返回的data对象并没有返回从网站上看到的JSON数据，而是得到了一个200响应。如果对data()对象使用dir方法，则会看到以下内容：

```
>>> dir(data)
['__attrs__', '__bool__','__class__','__delattr__','__dict__','__dir__','__doc__','__enter__', '__eq__', '__exit__', '__format__', '__ge__', '__getattribute__', '__getstate__', '__gt__', '__hash__', '__init__', '__init_subclass__', '__iter__', '__le__', '__lt__', '__module__', '__ne__', '__new__', '__nonzero__', '__reduce__', '__reduce_ex__', '__repr__', '__setattr__', '__setstate__', '__sizeof__', '__str__', '__subclasshook__', '__weakref__', '_content', '_content_consumed', '_next','apparent_encoding','close','connection','content', 'cookies', 'elapsed', 'encoding', 'headers', 'history', 'is_permanent_redirect', 'is_redirect', 'iter_content', 'iter_lines', 'json', 'links', 'next', 'ok', 'raise_for_status', 'raw', 'reason', 'request', 'status_code', 'text', 'url']
>>> data.json()
{'hello': 'world'}
>>> data.status_code
200
>>> data.reason
'OK'
>>> data.text
'{\n  "hello": "world"\n}\n'
>>> data.url
'http://127.0.0.1:5000/'
```

---

① my_flask_api.py脚本文件可以使用PyCharm运行，不要退出。交互式命令行程序可以借助Python命令行或Spyder中的IPython命令行执行，前提是要提前安装requests包。

由以上代码可见,借助dir()方法可以看到data对象的方法和属性。代码中查看了几个具体的方法和属性。毫无疑问,data对象的json()方法返回了从Web浏览器看到的JSON数据。status_code属性返回了Web请求的状态,本例返回的是200,代表是一个成功的请求。这里不对所有的状态代码进行一一讲解,相关内容可以在网上轻松查到。除了status_code属性外,本例中还给出了reason属性,该属性返回的内容可以与status_code属性返回内容放在一起。text属性返回的内容也可以借助使用的URL获取。很明显,我们可以通过requests包获得很多信息。

上述示例作为第一个示例,只演示了get请求,所设计的信息实际上不太有用,只是返回了一些简单的JSON数据。我们接下来要做的是创建一个API,通过编写代码将允许从一个小型电影数据库中查询(get)、创建(post)和删除(delete)信息。完整的代码如下所示:

```python
from flask import Flask
from flask_restful import reqparse, abort, Api, Resource

app = Flask(__name__)
api = Api(app)

film_dict = {
    '1': {'Name': 'Avengers: Infinity War', 'Year': 2018, 'Month': 'March'},
    '2': {'Name': 'Ant Man and the Wasp', 'Year': 2018, 'Month': 'August'},
}
def abort_if_todo_doesnt_exist(film_id):
    if film_id not in film_dict:
        abort(404, message="Film {} doesn't exist".format(film_id))


parser = reqparse.RequestParser()
parser.add_argument('name')
parser.add_argument('year')
parser.add_argument('month')

class Films(Resource):
    def get(self, film_id):
        abort_if_todo_doesnt_exist(film_id)
        return film_dict[film_id]

    def delete(self, film_id):
        abort_if_todo_doesnt_exist(film_id)
        del film_dict[film_id]
        return '', 204

    def put(self, film_id):
        args = parser.parse_args()
        task = {'Name': args['name'],
        'Year': args['year'],
```

```
                'Month': args['month']}
            film_dict[film_id] = task
            return task, 201

class FilmDict(Resource):
    def get(self):
        return film_dict

api.add_resource(FilmDict, '/films')
api.add_resource(Films, '/films/<film_id>')
if __name__ == '__main__':
    app.run(debug = True)
```

下面将对上述程序进行逐行说明。首先来看前两条语句,它们用于导入本例 API 程序用到的包。

```
from flask import Flask
from flask_restful import reqparse, abort, Api, Resource
```

这与前面 hello world 示例中导入的包类似。然而,本例从 flask-restful 包中导入的模块多了两个,即 reqparse 和 abort。reqparse 是一个参数解析器(argument parser),用于处理随 Web 请求发送到 API 的参数,如下所示:

```
parser = reqparse.RequestParser()
parser.add_argument('name')
parser.add_argument('year')
parser.add_argument('month')
```

上述代码创建了一个 RequestParser 对象,然后利用 add_argument() 方法添加 name、year 和 month 参数,当添加电影时将传递这些参数。由于添加之后还不允许获取参数,所以需要解析它们,在稍后的代码中有以下片段:

```
args = parser.parse_args()
```

该条语句的功能是从 RequestParser 解析参数,并将其存储在字典中。

导入的 abort 模块在自定义函数中使用,如果 film_id 不存在,则将使用该模块发送适当的消息。

```
def abort_if_todo_doesnt_exist(film_id):
    if film_id not in film_dict:
        abort(404, message = "Film {} doesn't exist".format(film_id))
```

上述代码中,film_id 作为函数的输入参数,如果传入的 film_id 在字典中不存在,则返回 404,并返回一个与不存在电影相关的自定义消息。该函数将会在代码中的许多地方用到。

接下来,考虑代码中定义的两个类:Films 类和 FilmDict 类。Films 类的定义如下:

```python
class Films(Resource):
    def get(self, film_id):
        abort_if_todo_doesnt_exist(film_id)
        return film_dict[film_id]

    def delete(self, film_id):
        abort_if_todo_doesnt_exist(film_id)
        del film_dict[film_id]
        return '', 204

    def put(self, film_id):
        args = parser.parse_args()
        task = {'Name': args['name'],
                'Year': args['year'],
                'Month': args['month']}
        film_dict[film_id] = task
        return task, 201
```

由上述代码可知,Films 类定义了 get()、delete() 和 put() 方法并可以实现相应功能,运用这些方法可以获取电影、删除电影或将电影放入字典中。需要注意的是,在 get() 方法和 delete() 方法中,首先使用 abort_if_film_doesnt_exist() 函数检查电影是否存在,并发送相应的错误消息和状态代码。实际上,可以在每个方法中都这样做,但如果使用了 20 个不同的方法,这将造成大量的重复代码。这段代码的实际操作非常简单,其使用了本书中介绍的字典相关方法。注意,对于 delete() 方法和 put() 方法,两者都返回了一个状态代码,delete() 方法同时还会返回一个空字符串,而 put() 方法会同时返回 task 字典内容。

```python
class FilmDict(Resource):
    def get(self):
        return film_dict
```

FilmDict 类只有一个 get() 方法,该方法返回整个电影数据库,但为什么要将其作为一个单独的类呢?这是因为我们将其添加到一个单独的 URL 中,如下面的代码片段所示:

```python
api.add_resource(FilmDict, '/films')
api.add_resource(Films, '/films/<film_id>')
```

现在,如果想基于 film_id 查询(get)、删除(delete)和更新(put)影片,可以通过影片的 URL 访问 FilmsDict 类,并需要通过 '/films/<film_id>' 传递参数。

```python
if __name__ == '__main__':
    app.run(debug = True)
```

和之前一样，我们在调试模式下运行应用程序。为了使其运行，将目录更改为代码所在的位置，将在 http://127.0.0.1:5000/ 上 API 可用的时间运行脚本文件，因此需要将网址导航到 http://127.0.0.1:5000/films，则会显示电影列表，如图 18.4 所示。

图 18.4　从浏览器获得所有电影的 API 显示

我们可以传递一个存在的电影 id，得到图 18.5 所示的结果。

图 18.5　通过浏览器查看 id 为 1 的电影的 API 显示

如果传递的电影 id 不存在，就会得到图 18.6 所示的消息。

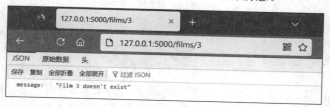

图 18.6　通过浏览器查看 id 为 3 的电影的 API 显示

现在，如果想要使用 API 添加或删除影片，可以使用 requests 库完成。接下来展示一些代码示例。

```
>>> import requests
>>> r = requests.get('http://127.0.0.1:5000/films')
>>> r.json()
{'1': {'Name': 'Avengers: Infinity War', 'Year': 2018, 'Month': 'March'},
'2': {'Name': 'Ant Man and the Wasp', 'Year': 2018, 'Month': 'August'}}
>>> r = requests.get('http://127.0.0.1:5000/films/1')
```

```
>>> r.json()
{'Name': 'Avengers: Infinity War', 'Year': 2018, 'Month': 'March'}
```

此时，可以使用 requests 从浏览器调用与之前相同的 URL。下面展示如何使用 API 进行添加和删除操作。

```
>>> r = requests.delete('http://127.0.0.1:5000/films/1')
>>> r.text
''
>>> r.status_code
204
```

正如上述代码所示，删除影片需要基于影片的 id，我们使用 delete() 方法，并将 URL 作为参数传入，代码中显示的 text 属性和 status_code 属性与 API 代码相匹配。现在，可以使用 put() 方法将删除的影片重新添加回去。

```
>>> r = requests.get('http://127.0.0.1:5000/films')
>>> r.json()
{'2': {'Name': 'Ant Man and the Wasp', 'Year': 2018, 'Month': 'August'}}
>>> r = requests.put('http://127.0.0.1:5000/films/1',
    data = {'name': 'Avengers: Infinity War',
    'year': 2018,
    'month': 'March'})
>>> r.json()
{'Name': 'Avengers: Infinity War', 'Year': '2018', 'Month': 'March'}
>>> r.status_code
201
>>> r = requests.get('http://127.0.0.1:5000/films')
>>> r.json()
{'2': {'Name': 'Ant Man and the Wasp', 'Year': 2018, 'Month': 'August'},
'1': {'Name': 'Avengers: Infinity War', 'Year': '2018', 'Month': 'March'}}
```

由上述代码可见，程序开始只有 id 为 2 的电影数据，后面使用 put() 方法添加数据（通过使用 data 参数添加 id 为 1 的数据）。在发送 put 请求时，以添加数据的形式与 API 中定义的变量名相匹配。最后，打开网址 http://127.0.0.1:5000/films 获取所有电影，以此检查电影是否被成功添加。可以看到，id 为 1 的电影已被成功添加。

到目前为止，我们编写的 API 都是公开的，如果想以某种方式进行保护，该怎么办呢？为此，需要在 API 上进行一些身份验证（authentication），下面将讨论一些可用的认证方式。

基本认证（basic authentication，Basic Auth）：Basic Auth 通过服务器询问用户名和密码（如个人社交媒体账户）来允许访问私有数据（如果数据在传输时没有加密，就会存在潜在的安全风险）。基本认证通过 HTTP 执行，并使用 SSL 加密，以保护用户名和密码的传输。注意，基本认证可以在没有 SSL 的情况下完成，但敏感数据将通过未加密的 HTTP 进行传输，这是一个极端的安全风险。Basic Auth 是一种简单的身份验证技术，它使脚本编码过程变得十分简单。但是,由于它依赖于使用用户名和密码访问 API 并管理与之关联的账户,所以这并不理想。这就像你把钥匙借给你的朋友,钥匙可以打开你家里和工作场所的锁。换句话说,如果你把自己的社交媒体用户名和密码交给脚本,这些脚本最终会在访问你的个人社交媒体账户方面超出想象!

API 密钥验证(API key authentication):API 密钥验证采用了密钥技术,它通过要求使用唯一的密钥访问 API 来克服 Basic Auth 的弱点。密钥通常是一长串字母和数字,与用户的登录信息(如用户名和密码)完全分离。因此,出于安全方面考虑,可以有意限制 API 密钥的获取,只提供对我们需要的数据和服务用户的访问,而不是授予对所有内容的访问。

OAuth 令牌(OAuth token):OAuth 是一个流行的标准,应用程序可以使用它为客户端应用程序提供使用 API 密钥验证原理操作的安全访问。OAuth 通过访问令牌而不是凭据(credentials)授权设备、API、服务器和应用程序。当使用 OAuth 对服务器连接进行身份验证时,从客户端应用程序(在本例中,是我们构建的 Python 脚本)发送身份验证请求到身份验证服务器。身份验证服务器生成 OAuth 令牌。令牌通过 HTTPS 返回给客户端应用程序,然后将其传递给 API 服务器。你可能会遇到要求你登录谷歌、Facebook 或 Twitter 账户的网站或应用程序,此时,谷歌、Facebook 或 Twitter 充当身份验证服务器。注意,身份验证服务器不要求是第三方服务器,但通常是与提供数据的服务器不同的服务器。

在本章最后一个例子中,我们将创建一个使用 Basic Auth 验证 Web 请求的 API。完整代码如下所示:

```python
from flask import Flask
from flask_restful import Resource, Api
from flask_httpauth import HTTPBasicAuth

app = Flask(__name__)
api = Api(app)
auth = HTTPBasicAuth()

USER_DATA = {
"admin": "atCh_5K}?g"
}

@auth.verify_password
def verify(username, password):
    if not (username and password):
        return False
    return USER_DATA.get(username) == password

class HelloWorld(Resource):
    @auth.login_required
    def get(self):
        return {"Hello": "World"}
```

```
api.add_resource(HelloWorld, '/hello_world')

if __name__ == '__main__':
    app.run(debug = True)
```

由于这段代码和之前的代码很相似,所以只看一下新添加的内容,其主要有两部分。

```
from flask import Flask
from flask_restful import Resource, Api
from flask_httpauth import HTTPBasicAuth

app = Flask(__name__)
api = Api(app)
auth = HTTPBasicAuth()

USER_DATA = {
"admin": "atCh_5K}?g"
}
```

上述代码中,需要从 flask_httpauth 包中导入 HTTPBasicAuth 模块(安装 flask_httpauth 包的方法可以参考本书前面的内容)。导入之后,创建了一个 HTTPBasicAuth() 对象,并将其赋值给 auth。随后创建了一个包含用户数据的字典 USER_DATA,将用户名和密码作为键和值。

```
@auth.verify_password
def verify(username, password):
    if not (username and password):
        return False
    return USER_DATA.get(username) == password

class HelloWorld(Resource):
    @auth.login_required
    def get(self):
        return {"Hello": "World"}
```

接下来的代码中,我们创建了一个 verify() 函数,它接收用户名(username)和密码(password)作为参数,如果使用 username 作为 get() 方法的输入参数获得的字典值与我们拥有的密码相同,则返回 True。注意,我们使用了字典的 get() 方法,所以不会出现键错误。verify() 函数使用 @auth.verify_password 装饰,因而可以使用 @auth.login_required 对 get() 方法进行装饰,这就意味着必须登录才能获得 HelloWorld 类的返回。资源被添加到端点/hello_world,可以按照前面示例中的方式运行 API。

在 API 运行之后,可以尝试利用 requests 库访问端点 http://127.0.0.1:5000/hello_world。

```
>>> r = requests.get('http://127.0.0.1:5000/hello_world') ①
>>> r.status_code
401
>>> r = requests.get('http://127.0.0.1:5000/hello_world',
auth = ('admin', 'atCh_5K}?g')) ②
>>> r.status_code
200
>>> r.json()
{'Hello': 'World'}
```

上述代码中,第①条 get()方法语句使用了标准的方法,该语句执行之后返回 401 状态代码,表示未获得授权。第②条 get()方法语句中,为了获得授权,将用户名和密码的元组传递给 auth 参数,成功访问端点并得到了预期的 JSON 响应。

本章小结

本章介绍了如何创建和访问 API。本章所有的例子都在本地机器上运行,Python 对于编写高质量 API 来说是一个很好的工具。如何访问 API 的示例对于那些希望使用不同数据源的人特别有用,因为 API 是一种可以让用户进行交互的常用解决方案。虽然 requests 库非常适合直接与 API 进行交互,但有些包将与 API 相关的复杂功能进行了封装。

CHAPTER 19

第 19 章 Python 网络爬虫

本书最后一章将介绍网络爬虫的概念。网络爬虫是自动从网页上抓取信息的过程。要掌握网络爬虫的知识,需要快速掌握以下几点内容:
- 超文本标记语言(HTML);
- 获取网页;
- 从网页上获取信息。

为了做到这一点,我们将使用 Python 创建自己的网站,并编写代码抓取网站上的信息。

19.1 HTML 简介

HTML 是超文本标记语言(hypertext markup language)的缩写,是创建网页的标准标记语言。人们在互联网上看到的东西本质上是由 HTML 语言构建的,HTML 文件告诉浏览器如何在网页上显示文本、图像和其他内容,其目的是描述网页内容的构建,而不是如何在 Web 浏览器中设置样式和渲染(render)。网页渲染可以使用 CSS(cascading style sheet,层叠样式表),HTML 页面可以链接到 CSS 文件,以获取关于颜色、字体和与页面渲染相关的其他信息。

HTML 是一种标记语言,因此在创建 HTML 内容时,需要把要显示的文本嵌入显示的页面中。借助 HTML 标签(tag)可以实现该功能,HTML 标签可以包含名称-值对(name-value pairs),也称作属性(attributes)。HTML 标签中的信息称为 HTML 元素。良好的 HTML 格式由开始标签和结束标签组成,在书写一个新标签之前,应该关闭之前的标签。

前面已经介绍了 HTML 的概念,后面将通过示例说明如何创建其中的元素,并展示如何设计一个页面。需要注意的是,当新建一个标签时需要在结束标签上加上一个正斜杠(/)。

下面介绍常用的 HTML 标签,让我们从 header 标签开始介绍[①]。

头部(header)

下面的语句是 header 标签的定义方法,其中< h1 >和</ h1 >是一对 header 标签[②]。

```
< h1 > This is how we define a header </ h1 >
```

在这里,< h1 >为开始标签:

```
< h1 >
```

</ h1 >为结束标签:

```
</ h1 >
```

段落(paragraph)

接下来,我们将展示如何定义 paragraph 标签:

```
< p > This is how we define a paragraph </ p >
```

定义(define)

下面,我们将展示如何定义 a 标签,并在其内部嵌入一个链接:

```
< a href = "https://www.google.com"> This is how we define a link </ a >
```

表格(table)

下一个 HTML 标签是 table 标签,该标签用于创建表格,这一标签比前面介绍的标签要复杂一些。

```
< table >
    < tr >
        < th > Name </ th >
        < th > Year </ th >
        < th > Month </ th >
    </ tr >
    < tr >
        < td > Avengers: Infinity War </ td >
        < td > 2018 </ td >
        < td > March </ td >
    </ tr >
    < tr >
        < td > Ant Man and the Wasp </ td >
        < td > 2018 </ td >
```

[①] 本章示例采用 VS Code 验证,排版与字体颜色都采用 VS Code 默认风格。

[②] header 标签总共有 6 个,即 h1、h2、h3、h4、h5 和 h6,从 h1 到 h6 对网页内容的重要性逐渐递减,h1 最重要,h6 最轻,类似于 Word 文档中的标题级别。此处应为 h1,原书写为 h,已更正。

```
        <td> August </td>
    </tr>
</table>
```

创建表格时首先要做的就是定义 table 标签:

```
<table></table>
```

接下来,我们需要使用 tr 标签定义表格的一行,并嵌套在 table 标签内。tr 标签定义如下:

```
<tr></tr>
```

上述示例中,我们使用 tr 标签 3 次,因此定义了 3 行表格。

在表格的每一行中,我们都定义了一些数据,其中使用了 th 标签和 td 标签。th 标签定义了表格中的表头单元格,而 td 标签定义了表格中的数据单元格。因此,本例中表格的标题是 Name、Year 和 Month,接下来是两行表格数据。

表头(thead)和表格主题(tbody)

我们可以向这个表中添加另外两个标签,即 thead 标签和 tbody 标签。在一个 table 里,可以使用 head 标签和 body 标签将表分为两部分,这两个标签的用法如下:

```
<table>
    <thead>
    <tr>
        <th> Name </th>
        <th> Year </th>
        <th> Month </th>
    </tr>
    </thead>
    <tbody>
        <tr>
            <td> Avengers: Infinity War </td>
            <td> 2018 </td>
            <td> March </td>
        </tr>
        <tr>
            <td> Ant Man and the Wasp </td>
            <td> 2018 </td>
            <td> August </td>
        </tr>
    </tbody>
</table>
```

分隔(div)

最后一个介绍的标签是 div 标签,该标签定义了 HTML 中的一节。因此,采用前面表的示例,我们在其中放置一个 div 标签。在 HTML 中使用 div 标签时,可以将整个一节的内容都放在 div 标签内部。

```
<div>
    <table>
        <thead>
            <tr>
                <th>Name</th>
                <th>Year</th>
                <th>Month</th>
            </tr>
        </thead>
        <tbody>
            <tr>
                <td>Avengers: Infinity War</td>
                <td>2018</td>
                <td>March</td>
            </tr>
            <tr>
                <td>Ant Man and the Wasp</td>
                <td>2018</td>
                <td>August</td>
            </tr>
        </tbody>
    </table>
</div>
```

HTML 属性（HTML attributes）

在介绍了几种标签的使用之后，现在讨论标签的属性。属性可以提供关于元素的一些附加信息。下面，我们将展示一些属性与前面所定义标签之间的关系。首先，我们将展示 a 标签的属性示例。

```
<a href='https://www.google.com">This is how we define a link</a>
```

上述语句中，href 是一个属性，用于指定链接目标的 URL。

我们还可以在标签中添加一个标题（title），此时 title 的值将显示为 tooltip 提示框（当将鼠标悬停在其上时会显示该提示信息）。

```
<p title="It will show when you hover over the text">This is how we define a paragraph</p>
```

id 属性和 class 属性

下面来看两个重要的属性，即 id 属性和 class 属性，这两个属性可以帮助我们在 HTML 中定位元素。元素的 id 属性在 HTML 文件中是唯一的，而 class 属性可以在多个元素中使用。下面，让我们通过带有 3 个 header 标签及其相关信息的 HTML 代码来展示其用法。

```
<!-- A unique element -->
< h1 id = 'myHeader'> MCU Films </h1 >
<!-- Multiple similar elements -->
< h2 class = 'film'> Avengers: Infinity War </h2 >
< p > Can Earth's mightiest heroes protect us from the threat of Thanos?</p >
< h2 class = 'film'> Ant Man and the Wasp </h2 >
< p > Following the events of Civil War will Scott Lang don the Ant Man suit again?</p >
< h2 class = 'film'> Captain Marvel </h2 >
< p > Plot Unknown.</p >
```

上述脚本中有一个带有 id 属性的 header 标签，这一属性具有唯一性，用于对 header 标签进行标识。其余的 header 标签都有 class 属性，并且每个 class 属性的值都为 "film"。

以上这些内容仅仅是对 HTML 的简要介绍，对本章的其余部分有所帮助。要对 HTML 语言进行全面掌握还需要进行深入研究，如果您感兴趣，可以在网上找到很多关于 HTML 的学习资源。

19.2 网页抓取

在介绍了 HTML 及其工作原理之后，现在介绍如何使用 Python 获取网页数据。当要进行网页抓取时，我们通常会想到从网站获取和处理数据。实际上，这可以分为两个不同的过程：网络爬取（web crawling）和网页抓取（web scraping）。

（1）网络爬取是从一个或多个 URL 网页获取数据的过程，这些 URL 可以通过遍历 HTML 网站获得。例如，某个网站首页有很多指向其他页面的链接，如果你想从所有链接中获取所有信息，便可以通过编程遍历所有链接，然后访问所有相关页面，并将其进行存储。

（2）网页抓取是从想要查询的页面获取信息的过程。而对网络爬取来说，需要首先抓取页面以获取想要遍历的链接。在抓取页面时，将以编程方式从页面中获取信息，并且要对获取的信息进行存储和数据处理。

网页数据抓取在整个网络爬取过程中扮演着重要的角色，也被认为是网络爬取和与网页抓取的结合。在抓取页面时并没有限制抓取的行为准则，然而，当人们试图从网站获取数据时，就需要注意一些重要的规则。

- 检查是否被允许从指定的网站获取和使用数据：虽然您可能认为网站上的任何数据都是可以获取的，但事实并非总是如此。注意查看网站的使用条款，因为虽然它们可能无法阻止您通过网络抓取获取数据，但如果在研究中使用了网站上抓取的数据，当对方看到时可能会引发问题，所以要小心。从网上获取数据的合法性问题非常重要，如果存在疑问就需要寻求法律建议。本书中采用我们在本地创建的网页作为示例，并从中获取数据，从而介绍网页数据抓取方法。

- 检查是否有公平使用政策：如果方式合适，有些网站很乐意让人抓取数据。这是为什么呢？其实，每当你访问网站时，都在为该网站提供流量。但是，如果通过计算机去访问时，便可以非常快速地向站点发送大量访问请求，这可能会给站点带来问题。如果网站监测到该行为，访问者的 IP 地址可能会被阻止，也就意味着无法再去访问该网站。因此，需要考虑代码访问网站的频率，考虑访问的必要性与适当性，并考虑网站是否允许这样做。对于调用单个 URL 的代码，主要考虑该代码的运行频率，然而，如果代码针对多个 URL 网页进行数据爬取，则需要确保代码以适当的速度运行。要做到这一点，需要考虑增加时间延迟，以确保不会向站点发送太多的访问请求。
- robots.txt：对于上面的问题，通过访问网站上的 url/robots.txt 文件可以获得有关网络爬虫的说明，该文件说明什么可以做什么不能做。如果存在该文件，就应该认真阅读，以了解网络爬虫许可的范围。如果没有 robots.txt 文件，就没有关于该网站执行网络爬取的具体说明，但这也不能说明可以随意从网站上爬取任何内容。

总之，进行网络抓取时要小心，如果网站提供了可用的应用程序编程接口（API），就应该使用它。如果不确定是否有 API 接口，需要请求获得适当的建议。在开始网站爬取和数据抓取之前，您需要了解页面内容。在数据爬取时，不能只是让 Python 获取所需的数据，而是要告诉 Python 如何查找所需数据。因此，需要了解 HTML 的工作原理，并知道到要查找数据的位置。

要获取网页数据，可以通过两种方法：通过 Web 浏览器；保存网页并查找。

通过 Web 浏览器

可以借助 Web 浏览器检查 HTML 如何引用了页面的元素，然而，查看网页内容的方式决定于浏览器本身。不管怎样，都需要选择页面的元素并检查相应的 HTML，然后显示 HTML 引用该元素的内容。浏览器不同，检查其页面显示内容的方式也不同，因此需要参阅所使用浏览器的帮助文档。

保存网页并查找

可以保存网页，然后搜索指定文本的名称或值，并以同样的方式确定 HTML 引用该元素的内容。最终，需要试着搞清楚 HTML 如何实现网页数据的展示。这不是精准的科学问题，因为可以用不同的方式编写 HTML。关键是理解 HTML 的定义以及它们是如何组合在一起的，然后使用它来理解需要在 HTML 页面中访问什么。对于任何一段代码，都需要提前规划，计划好如何从 HTML 获取数据，从而实现 HTML 的解析。

在此，我们重新回到一开始提到的内容，即网络爬取和网页抓取的问题。网络爬取是从一个或多个 URL 地址访问数据的过程。我们需要一种机制来实现这一功能，幸运的是，发出 Web 请求的方式与第 18 章中访问 API 端点的方式相同，因此我们可以使用 requests 库。稍后我们将在自己建立的网页中进行演示。

具有获取数据的机制是很好的，但还需要处理得到的数据，因此我们需要利用一个可以实现该功能的 Python 库。因为 Python 有很多可用的选项，而且 Python 语言也在不断发展,所以本书目的并不是要找到处理 HTML 的最佳包。相反,我们将介绍一个专门的 HTML 解析器,即 BeautifulSoup。

BeautifulSoup

BeautifulSoup 不仅是一个 HTML 解析器,还可以通过使用 14.3 节介绍的 lxml 库来解析 XML。BeautifulSoup 工作时会用到其他解析器。因此,要运行 BeautifulSoup,需要先执行以下语句:

```
>>> from bs4 import BeautifulSoup
>>> BeautifulSoup(content_to_be_parsed, "parser_name")
```

上述语句中,content_to_be_parsed 是来自站点的内容,可以使用前面提到的 requests 包获得;"parser_name"是要使用的解析器的名称。下面是常用的 4 种解析器:

- **html.parser**:Python 默认的 HTML 解析器,速度很快,能较好地兼容 HTML3.2.2。
- **lxml**:基于 Python 的 lxml 库构建,兼容性好,速度很快。
- **lxml-xml 或 xml**:基于 lxml 构建,与 lxml 类似,但也是 BeautifulSoup 唯一可用的 xml 解析器。因此,虽然我们介绍了如何使用 lxml 解析 XML,但也可以在 BeautifulSoup 中执行同样的操作。
- **htmllib5**:基于 Python 的 html5lib 库构建,速度非常慢,但是它兼容性非常好。htmllib5 通过 Web 浏览器创建有效 HTML5 的方式解析页面。

在本节下面的内容中,我们将主要采用 html.parser。可以执行如下命令[①]创建 BeautifulSoup 解析器:

```
>>> import requests
>>> from bs4 import BeautifulSoup
>>> url = "some_url"
>>> r = requests.get(url)
>>> response_text = r.text
>>> soup = BeautifulSoup(response_text, "html.parser")
```

这样就可以把 HTML 转换为可以访问其中元素的格式。下面将介绍可用的方法:

```
>>> text = '<b class = "boldest">This is bold</b>'
>>> soup = BeautifulSoup(text,"html.parser")
>>> soup
<b class = "boldest">This is bold</b>
```

可以利用下述方法访问该标签:

```
>>> tag = soup.b
>>> tag
<b class = "boldest">This is bold</b>
```

[①] 下述代码中"some_url"不是合法的网页地址,运行 get()方法报错,用于示例作用。

如果有多个 b 标签,使用 soup.b 将只能返回第一个,示例如下:

```
>>> text = """<b class = "boldest">This is bold</b>
... <b class = "boldest">This also is bold</b>"""
>>> soup = BeautifulSoup(text,"html.parser")
>>> tag = soup.b
>>> tag
<b class = "boldest">This is bold</b>
```

因此,只能取回第一个 b 标签,而不是取回所有的 b 标签。标签本身有一个名称,可以通过如下方式进行访问:

```
>>> tag.name
'b'
```

标签还具有一个属性字典,可按如下方式访问:

```
>>> tag.attrs
{'class': ['boldest']}
>>> tag["class"]
['boldest']
```

现在,让我们看一个更复杂的示例。如果查看类似表的内容,可以按如下方式解析它:

```
>>> text = """<table>
...     <tr>
...         <th>Name</th>
...         <th>Year</th>
...         <th>Month</th>
...     </tr>
...     <tr>
...         <td>Avengers: Infinity War</td>
...         <td>2018</td>
...         <td>March</td>
...     </tr>
...     <tr>
...         <td>Ant Man and the Wasp</td>
...         <td>2018</td>
...         <td>August</td>
...     </tr>
... </table>"""
>>> text
'<table>\n... <tr>\n... <th>Name</th>\n... <th>Year</th>\n... <th>Month</th>\n... </tr>\n... <tr>\n... <td>Avengers: Infinity War</td>\n... <td>2018</td>\n... <td>March</td>\n... </tr>\n... <tr>\n... <td>Ant Man and the Wasp</td>\n... <td>2018</td>\n... <td>August</td>\n... </tr>\n... </table>'
```

```
>>> soup = BeautifulSoup(text, "html.parser")
>>> soup
<table>
...<tr>
...<th>Name</th>
...<th>Year</th>
...<th>Month</th>
...</tr>
...<tr>
...<td>Avengers: Infinity War</td>
...<td>2018</td>
...<td>March</td>
...</tr>
...<tr>
...<td>Ant Man and the Wasp</td>
...<td>2018</td>
...<td>August</td>
...</tr>
...</table>
```

现在,我们可以通过遍历 HTML 的树结构来访问表的元素。

```
>>> soup.table
<table>
<tr>
<th>Name</th>
<th>Year</th>
<th>Month</th>
</tr>
<tr>
<td>Avengers: Infinity War</td>
<td>2018</td>
<td>March</td>
</tr>
<tr>
<td>Ant Man and the Wasp</td>
<td>2018</td>
<td>August</td>
</tr>
</table>
>>> soup.table.tr
<tr>
<th>Name</th>
<th>Year</th>
<th>Month</th>
</tr>
>>> soup.table.tr.th
<th>Name</th>
```

现在，在每个示例中，我们都得到了要查找标签的第一个实例。假设我们想查找表中所有的 tr 标签，可以使用 find_all() 方法，如下所示：

```
>>> soup.find_all("tr")
[<tr>
<th>Name</th>
<th>Year</th>
<th>Month</th>
</tr>, <tr>
<td>Avengers: Infinity War</td>
<td>2018</td>
<td>March</td>
</tr>, <tr>
<td>Ant Man and the Wasp</td>
<td>2018</td>
<td>August</td>
</tr>]
```

如上所示，借助 find_all() 方法，我们得到了所有 tr 标签的列表。同样，我们可以使用相同的方法获取 td 标签的列表。

```
>>> tags = soup.find_all("td")
>>> tags
[<td>Avengers: Infinity War</td>, <td>2018</td>, <td>March</td>, <td>Ant Man and the Wasp</td>, <td>2018</td>, <td>August</td>]
```

因此，对于数据获取来说，可以回顾并考虑原始的表格，使用 find_all() 方法获取所有 tr 标签，然后对标签列表进行循环，就可以获取每个 td 标签数据，示例代码如下：

```
>>> table_rows = soup.find_all("tr")
>>> table_rows
[<tr>
<th>Name</th>
<th>Year</th>
<th>Month</th>
</tr>, <tr>
<td>Avengers: Infinity War</td>
<td>2018</td>
<td>March</td>
</tr>, <tr>
<td>Ant Man and the Wasp</td>
<td>2018</td>
<td>August</td>
</tr>]
>>> headers = []
>>> content = []
>>> for tr in table_rows:
...     header_tags = tr.find_all("th")
```

```
...         if len(header_tags) > 0:
...             for ht in header_tags:
...                 headers.append(ht.text)
...         else:
...             row = []
...             row_tags = tr.find_all("td")
...             for rt in row_tags:
...                 row.append(rt.text)
...             content.append(row)
...
>>> headers
['Name', 'Year', 'Month']
>>> content
[['Avengers: Infinity War', '2018', 'March'], ['Ant Man and the Wasp', '2018', 'August']]
>>>
```

上述代码对外层 tr 标签标进行循环以获取每一行，然后查找 th 标签，如果找到 th 标签，就找到了表的标题。如果没有找到，就去获取 td 标签，并将它们与列表的标题或内容相关联。需要注意是，我们知道数据的结构，因为我们已经检查了 HTML，所以在构建解析方案时就已经知道能获取什么内容。

至此，我们介绍了 find_all() 方法应用于一个表的情况。但是，如果有两个表该怎样处理呢？此时，每个要访问的表都有一个特定的 id，我们可以使用 find() 方法进行访问，示例代码如下：

```
>>> text = """<table id = "unique_table">
...     <tr><th> Name </th><th> Year </th><th> Month </th></tr><tr>
...     <td> Avengers: Infinity War </td><td> 2018 </td>" \
...     "<td> March </td></tr><tr>
...     <td> Ant Man and the Wasp </td><td> 2018
...     </td><td> August </td></tr>\n</table>" \
...     "<table id = "second_table">
...     <tr><th> Name </th><th> Year </th><th> Month </th></tr><tr>
...     <td> Avengers: End Game </td><td> 2019 </td>" \
...     "<td> April </td></tr>
...     <tr><td> Spider-man: Far from home </td>
...     <td> 2019 </td><td> June </td></tr>\n</table>
...     <table id = "other_table"><tr><th> Name </th>"""
>>>
>>> soup = BeautifulSoup(text, "html.parser")
>>> soup
<table id = "unique_table">
<tr><th> Name </th><th> Year </th><th> Month </th></tr><tr>
<td> Avengers: Infinity War </td><td> 2018 </td>"     "<td> March </td></tr><tr>
<td> Ant Man and the Wasp </td><td> 2018
    </td><td> August </td></tr>
</table>"     "<table id = "second_table">
<tr><th> Name </th><th> Year </th><th> Month </th></tr><tr>
```

```
<td>Avengers: End Game</td><td>2019</td>"        "<td>April</td></tr>
<tr><td>Spider-man: Far from home</td>
<td>2019</td><td>June</td></tr>
</table>
<table id="other_table"><tr><th>Name</th></tr></table>
>>> table = soup.find("table", id="second_table")
>>> table
<table id="second_table">
<tr><th>Name</th><th>Year</th><th>Month</th></tr><tr>
<td>Avengers: End Game</td><td>2019</td>"        "<td>April</td></tr>
<tr><td>Spider-man: Far from home</td>
<td>2019</td><td>June</td></tr>
</table>
```

我们可以采用之前的方式从该表中获取数据,但是也可以再次利用 find_all() 方法查找特定元素中的文本。

```
>>> table_rows = table.find_all("tr")
>>> table_rows
[<tr><th>Name</th><th>Year</th><th>Month</th></tr>, <tr>
<td>Avengers: End Game</td><td>2019</td>"        "<td>April</td></tr>, <tr><td>
Spider-man: Far from home</td>
<td>2019</td><td>June</td></tr>]
>>> headers = []
>>> content = []
>>> for tr in table_rows:
...     header_tags = tr.find_all("th")
...     if len(header_tags) > 0:
...         for ht in header_tags:
...             headers.append(ht.text)
...     else:
...         row = []
...         row_tags = tr.find_all("td")
...         for rt in row_tags:
...             row.append(rt.text)
...         content.append(row)
...
>>> headers
['Name', 'Year', 'Month']
>>> content
[['Avengers: End Game', '2019', 'April'], ['Spider-man: Far from home', '2019', 'June']]
```

上述示例表明,当有多个表时,我们可以从一个特定的表中获取信息,这实际上取决于表所具有的 id 属性,这种方式使得对表的访问过程更加容易。

至此,我们已经知道如何处理 HTML,下一步就要从网站获取数据,然后使用 Python 解析数据。本书将在本地建立自己的网站,从中获取数据并分析结果。前面已经介绍了获取和处理数据相关的库,下面看一下如何创建网站。

如第 18 章所述，我们将使用 flask 包创建一个简单的网站，在本地运行该网站，然后从中抓取数据。下面将通过一个简单的 Hello World 例子说明工作过程。首先，创建一个名为 my_flask_website.py 的文件，并在文件中写入以下代码：

```python
from flask import Flask
app = Flask(__name__)

@app.route('/')
def hello_world():
    return 'Hello World'

if __name__ == '__main__':
    app.run()
```

如果回想一下第 18 章，我们在这里看到的是之前创建 API 的简化版本。上述代码首先从 flask 包中导入 Flask，然后创建一个自己的 API 程序。与创建类的 API 程序不同，我们只定义了一个 hello_world() 函数，该函数返回字符串 Hello World，如下所示：

```python
@app.route('/')
def hello_world():
    return 'Hello World'
```

我们再次使用如下语法运行 API 程序：

```python
if __name__ == '__main__':
    app.run(debug = True)
```

与构建的 API 一样，如果我们打开终端或命令提示符，转到文件所在目录位置，然后使用 Python my_flask_website.py 命令运行代码，就会得到如图 19.1 所示的页面。

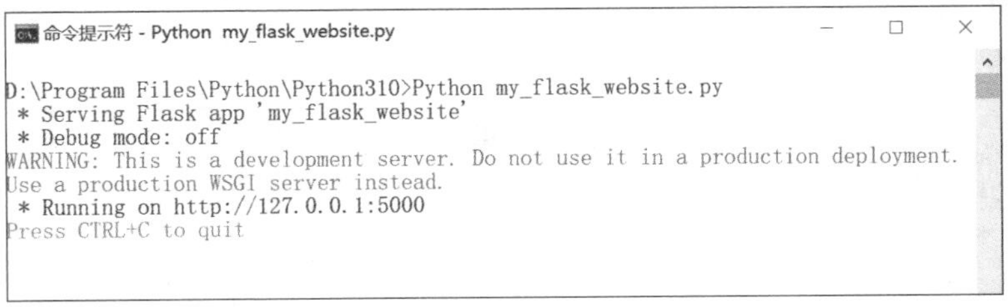

```
D:\Program Files\Python\Python310>Python my_flask_website.py
 * Serving Flask app 'my_flask_website'
 * Debug mode: off
WARNING: This is a development server. Do not use it in a production deployment.
Use a production WSGI server instead.
 * Running on http://127.0.0.1:5000
Press CTRL+C to quit
```

图 19.1　从终端显示网站

如果在 Web 浏览器上访问该地址，将看到如图 19.2 所示的页面。

上述代码中，有一段代码以前没有介绍过，即 @ app. route，它是一种装饰器（decorator）。本例中，该装饰器的目的是将网址绑定到函数，如下所示：

```python
@app.route('/')
def hello_world():
    return 'Hello World'
```

图 19.2 从浏览器显示网站

我们要做的是将 http://127.0.0.1:5000/映射到 hello_world()函数。当调用该 URL 地址时,将执行 hello_world()函数,并显示结果。这是装饰器的特定用法,通常情况下,装饰器是可以将函数作为参数的函数。示例演示是最好的讲解方法,下面用一个修饰器修饰 hello_world()函数,使字符串全部小写。示例代码如下[①]:

```
def make_lowercase(function):
    def wrapper():
        func = function()
        lowecase = func.lower()
        return lowercase

    return wrapper
```

此函数所做的是将 function 作为参数,在 return 语句中运行 wrapper()函数。wrapper()函数执行时将运行 function()函数,通过执行其中的 lower()方法使其变为小写并返回该值。将该函数加入 my_flask_website.py 代码中,如下所示:

```
from flask import Flask
app = Flask(__name__)

def make_lowercase(function):
    def wrapper():
        func = function()
        lowercase = func.lower()
        return lowercase

    return wrapper

@app.route('/')
@make_lowercase
def hello_world():
    return 'Hello World'
```

---

① 原书中将函数名写为 make_uppercase,实际应为 make_lowercase,已更正。

```
if __name__ == '__main__':
    app.run()
```

执行上述代码,在 Web 浏览器上访问 http://127.0.0.1:5000/,结果如图 19.3 所示。

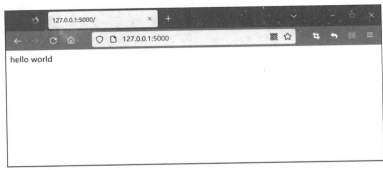

图 19.3　使用 make_lowercase() 函数和装饰器之后的网页显示

为简单起见,回到最初的 my_flask_website.py 文件,该文件中没有添加 make_lowercase() 函数和装饰器。启动服务器并运行我们的网站,可以通过 requests 库中的 get() 方法从网站中获取数据。示例代码如下:

```
>>> import requests
>>>
>>> r = requests.get('http://127.0.0.1:5000/')
>>> r.text
'Hello World'
```

注意,上述代码中我们查看的是 text 属性。此处没有用 json() 方法获取 JSON 数据是因为我们网站的内容不是 JSON 格式。这些看起来很容易操作,但由于没有采用图 19.4 所示的 HTML 格式,因此处理数据并不具有挑战性。

图 19.4　HTML 网页显示

我们只需修改 Flask 应用程序中的代码就可以很容易地实现这一点,将 Hello World 输出形式修改为 HTML 格式,示例代码如下:

```
from flask import Flask
app = Flask(__name__)

@app.route('/')
def hello_world():
    return '<h>Hello World</h>'

if __name__ == '__main__':
    app.run()
```

运行上述Flask应用程序,可以看到Web浏览器上的内容与我们之前看到的很相似,看不出有什么变化。但如果我们运行代码重新利用get()方法从网页获取数据,此时会得到以下结果:

```
>>> import requests
>>>
>>> r = requests.get('http://127.0.0.1:5000/')
>>> r.text
'<h>Hello World</h>'
```

从代码执行结果可见,text属性的内容发生了变化,变成了HTML格式数据,加上了h标签。

让我们再次修改代码,并将h标签更改为h1标签,修改之后Flask应用程序代码如下所示:

```
from flask import Flask
app = Flask(__name__)

@app.route('/')
def hello_world():
    return '<h1>Hello World</h1>'

if __name__ == '__main__':
    app.run()
```

运行上述Flask应用程序,在Web浏览器上访问http://127.0.0.1:5000/,可以看到如图19.5所示显示结果。

再次在该网站上运行相同的requests代码,将返回h1标签,如下所示:

```
>>> import requests
>>>
>>> r = requests.get('http://127.0.0.1:5000/')
>>> r.text
'<h1>Hello World</h1>'
```

至此,我们的网站已经可以运行了。下面,添加一些更难解析的内容,并创建一个可以通过编程获得的表。为此,我们将向Flask应用程序添加一个新路由(route),并添加

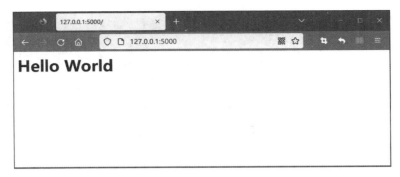

图 19.5　带 h1 标签的网站显示

一个 HTML 表。为此，我们将利用 Seaborn 库的 tips 数据集中的一些现有数据。

```
>>> import seaborn as sns
>>> tips = sns.load_dataset("tips")
>>> tips.head().to_html()
'<table border="1" class="dataframe">\n<thead>\n    <tr style="text-align:
right;">\n      <th></th>\n      <th>total_bill</th>\n      <th>tip</th>\n
<th>sex</th>\n      <th>smoker</th>\n      <th>day</th>\n
<th>time</th>\n      <th>size</th>\n    </tr>\n  </thead>\n  <tbody>\n
<tr>\n      <th>0</th>\n      <td>16.99</td>\n      <td>1.01</td>\n
<td>Female</td>\n      <td>No</td>\n      <td>Sun</td>\n
<td>Dinner</td>\n      <td>2</td>\n    </tr>\n    <tr>\n      <th>1</th>\n
<td>10.34</td>\n      <td>1.66</td>\n      <td>Male</td>\n
<td>No</td>\n      <td>Sun</td>\n      <td>Dinner</td>\n      <td>3</td>\n
</tr>\n    <tr>\n      <th>2</th>\n      <td>21.01</td>\n
<td>3.50</td>\n      <td>Male</td>\n      <td>No</td>\n      <td>Sun</td>\n
<td>Dinner</td>\n      <td>3</td>\n    </tr>\n    <tr>\n      <th>3</th>\n
<td>23.68</td>\n      <td>3.31</td>\n      <td>Male</td>\n
<td>No</td>\n      <td>Sun</td>\n      <td>Dinner</td>\n      <td>2</td>\n
</tr>\n    <tr>\n      <th>4</th>\n      <td>24.59</td>\n
<td>3.61</td>\n      <td>Female</td>\n      <td>No</td>\n
<td>Sun</td>\n      <td>Dinner</td>\n      <td>4</td>\n    </tr>\n
</tbody>\n</table>'
```

上述代码中，首先导入 tips 数据集，然后使用 Pandas 库的 to_html() 方法，该方法获取 DataFrame 并返回可以放在网站上的 HTML 数据。如果回顾前面关于表的示例，我们希望向表中添加一个 id，以允许对表的访问。可以向 to_html() 方法传递 table_id 参数，并为表设置一个期望的名称。可以将表的名称设置为 tips，以便后续使用该表，示例代码如下：

```
>>> import seaborn as sns
>>> tips = sns.load_dataset("tips")
>>> tips.head().to_html(table_id='tips')
'<table border="1" class="dataframe" id="tips">\n<thead>\n    <tr
style="text-align: right;">\n      <th></th>\n      <th>total_bill</th>\n
<th>tip</th>\n      <th>sex</th>\n      <th>smoker</th>\n
<th>day</th>\n      <th>time</th>\n      <th>size</th>\n    </tr>\n
```

```
</thead>\n  <tbody>\n    <tr>\n      <th>0</th>\n      <td>16.99</td>\n
<td>1.01</td>\n      <td>Female</td>\n      <td>No</td>\n
<td>Sun</td>\n      <td>Dinner</td>\n      <td>2</td>\n    </tr>\n    <tr>\n
<th>1</th>\n      <td>10.34</td>\n      <td>1.66</td>\n      <td>Male</td>\n
<td>No</td>\n      <td>Sun</td>\n      <td>Dinner</td>\n      <td>3</td>\n
</tr>\n    <tr>\n      <th>2</th>\n      <td>21.01</td>\n
<td>3.50</td>\n      <td>Male</td>\n      <td>No</td>\n      <td>Sun</td>\n
<td>Dinner</td>\n      <td>3</td>\n    </tr>\n    <tr>\n      <th>3</th>\n
<td>23.68</td>\n      <td>3.31</td>\n      <td>Male</td>\n
<td>No</td>\n      <td>Sun</td>\n      <td>Dinner</td>\n      <td>2</td>\n
</tr>\n    <tr>\n      <th>4</th>\n      <td>24.59</td>\n
<td>3.61</td>\n      <td>Female</td>\n      <td>No</td>\n
<td>Sun</td>\n      <td>Dinner</td>\n      <td>4</td>\n    </tr>\n
</tbody>\n</table>'
```

从上述代码可以看到,我们添加了名为 tips 的 id 属性。下一步,将其添加到网站,可以按如下方式修改代码:

```python
from flask import Flask
import seaborn as sns
app = Flask(__name__)

tips = sns.load_dataset("tips")

@app.route('/')
def hello_world():
    return '<h1>Hello World</h1>'

@app.route('/table')
def table_view():
    return tips.head(20).to_html(table_id='tips')

if __name__ == '__main__':
    app.run(debug=True)
```

上述代码的不同之处在于,我们导入了 Seaborn 库,并加载了 tips 数据集。为了显示其中的数据,创建了另一个名为 table_view 的函数,并在其中返回 DataFrame 的 20 行数据,接着将其转换为 id 为 tips 的 HTML 格式。随后,定义了一个装饰器并将此路由定义为/table,这意味着当转到 http://127.0.0.1:5000/table 的网址时,我们将看到该函数的返回结果。运行该代码启动服务器,并转到上述 URL 网址,可以在浏览器上看到如图 19.6 所示的结果。

现在,我们可以看到要显示的表,但看起来不太美观,可以使用 Pandas 附带的一些选项来进行定制显示。首先,可以从表中删除索引,因为该索引通常不会在网站上显示。接下来,可以将表格标题居中,并使边框更加突出。修改对应的 Flask 应用程序,示例代码如下所示:

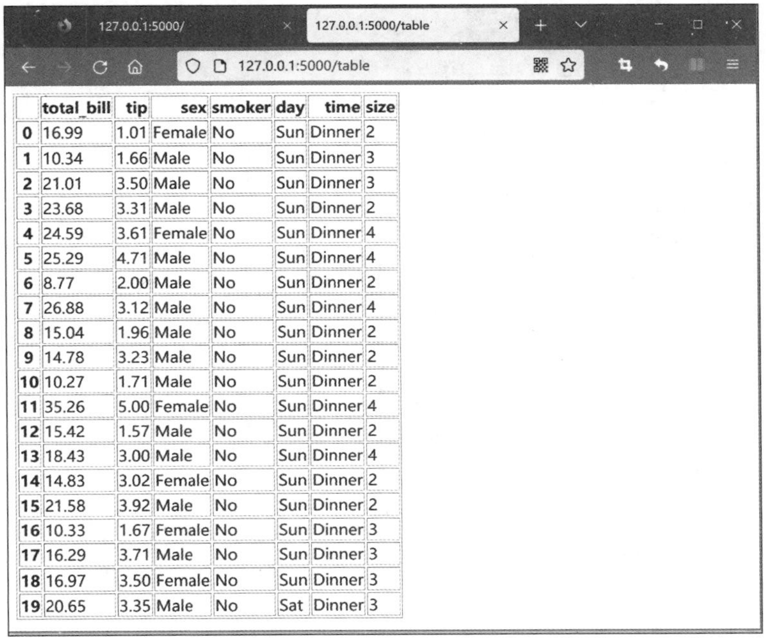

图 19.6 通过浏览器显示网站中的表

```
from flask import Flask
import seaborn as sns
app = Flask(__name__)

tips = sns.load_dataset("tips")

@app.route('/')
def hello_world():
    return '< h1 > Hello World </h1 >'

@app.route('/table')
def table_view():
    return tips.head(20).to_html(table_id = 'tips', border = 6,
index = False, justify = 'center')

if __name__ == '__main__':
    app.run(debug = True)
```

代码修改后,重新运行,浏览器显示结果如图 19.7 所示。

如果使用前面提到的一些内容,可以在一段中添加标题和有关网站的一些信息。为此,可以使用 h1 和 p 标签分别创建一个页眉和段落,并显示所有内容都属于一个 div 标签,这样它就类似于在生产网页上看到的内容。Flask 应用程序如下所示:

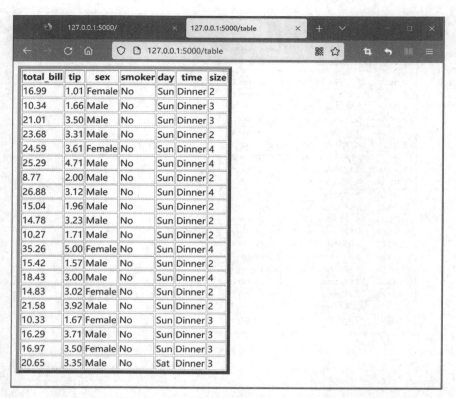

图 19.7 采用定制设置网站中表的显示

```
from flask import Flask
import seaborn as sns
app = Flask(__name__)

tips = sns.load_dataset("tips")

@app.route('/')
def hello_world():
    return '<h1>Hello World</h1>'

@app.route('/table')
def table_view():
    html = '<div><h1>Table of tips data</h1>' + \
        '<p>This table contains data from the seaborn tips dataset</p>' + \
        tips.head(20).to_html(table_id='tips', border=6,
        index=False, justify='center') + '</div>'
    return html

if __name__ == '__main__':
    app.run(debug=True)
```

代码按以上修改,重新运行,浏览器显示结果如图19.8所示。

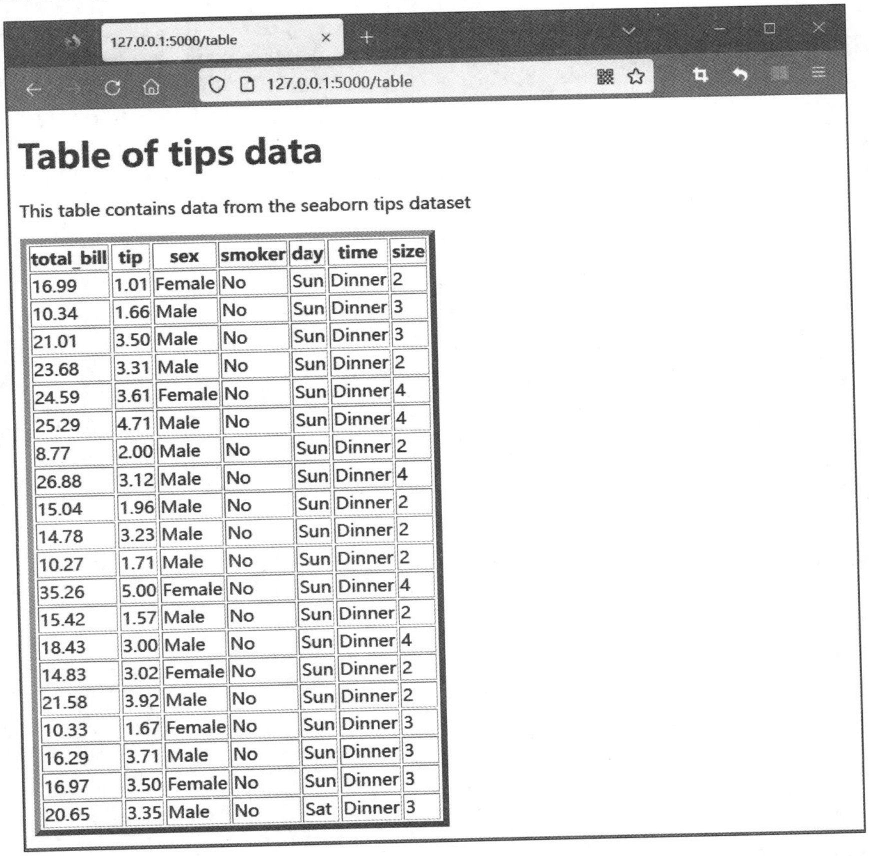

图19.8 带有头部和段落设置的网站中表的显示

至此,进行网页抓取的网站已经建好,下面就要抓取其中的内容。可以使用 requests 库的方法来获取 HTML 内容,示例代码如下:

```
>>> import requests
>>>
>>> r = requests.get('http://127.0.0.1:5000/table')
>>> r.text
'<div><h1> Table of tips data </h1><p> This table contains data from the seaborn tips
dataset </p><table border = "6" class = "dataframe" id = "tips">\n    <thead>\n    <tr
style = "text-align: center;">\n      <th> total_bill </th>\n      <th> tip </th>\n
<th> sex </th>\n      <th> smoker </th>\n      <th> day </th>\n      <th> time </th>\n
<th> size </th>\n    </tr>\n  </thead>\n  <tbody>\n    <tr>\n      <td> 16.99 </td>\n
<td> 1.01 </td>\n      <td> Female </td>\n      <td> No </td>\n      <td> Sun </td>\n
<td> Dinner </td>\n      <td> 2 </td>\n    </tr>\n    <tr>\n      <td> 10.34 </td>\n
<td> 1.66 </td>\n      <td> Male </td>\n      <td> No </td>\n      <td> Sun </td>\n
<td> Dinner </td>\n      <td> 3 </td>\n    </tr>\n    <tr>\n      <td> 21.01 </td>\n
<td> 3.50 </td>\n      <td> Male </td>\n      <td> No </td>\n      <td> Sun </td>\n
```

```
  < td > Dinner </ td >\n        < td > 3 </ td >\n          </ tr >\n         < tr >\n          < td > 23.68 </ td >\n
  < td > 3.31 </ td >\n          < td > Male </ td >\n       < td > No </ td >\n      < tr >\n            < td > Sun </ td >\n
  < td > Dinner </ td >\n        < td > 2 </ td >\n          </ tr >\n         < tr >\n          < td > 24.59 </ td >\n
  < td > 3.61 </ td >\n          < td > Female </ td >\n     < td > No </ td >\n      < tr >\n            < td > Sun </ td >\n
  < td > Dinner </ td >\n        < td > 4 </ td >\n          </ tr >\n         < tr >\n          < td > 25.29 </ td >\n
  < td > 4.71 </ td >\n          < td > Male </ td >\n       < td > No </ td >\n      < tr >\n            < td > Sun </ td >\n
  < td > Dinner </ td >\n        < td > 4 </ td >\n          </ tr >\n         < tr >\n          < td > 8.77 </ td >\n
  < td > 2.00 </ td >\n          < td > Male </ td >\n       < td > No </ td >\n      < tr >\n            < td > Sun </ td >\n
  < td > Dinner </ td >\n        < td > 2 </ td >\n          </ tr >\n         < tr >\n          < td > 26.88 </ td >\n
  < td > 3.12 </ td >\n          < td > Male </ td >\n       < td > No </ td >\n      < tr >\n            < td > Sun </ td >\n
  < td > Dinner </ td >\n        < td > 4 </ td >\n          </ tr >\n         < tr >\n          < td > 15.04 </ td >\n
  < td > 1.96 </ td >\n          < td > Male </ td >\n       < td > No </ td >\n      < tr >\n            < td > Sun </ td >\n
  < td > Dinner </ td >\n        < td > 2 </ td >\n          </ tr >\n         < tr >\n          < td > 14.78 </ td >\n
  < td > 3.23 </ td >\n          < td > Male </ td >\n       < td > No </ td >\n      < tr >\n            < td > Sun </ td >\n
  < td > Dinner </ td >\n        < td > 2 </ td >\n          </ tr >\n         < tr >\n          < td > 10.27 </ td >\n
  < td > 1.71 </ td >\n          < td > Male </ td >\n       < td > No </ td >\n      < tr >\n            < td > Sun </ td >\n
  < td > Dinner </ td >\n        < td > 2 </ td >\n          </ tr >\n         < tr >\n          < td > 35.26 </ td >\n
  < td > 5.00 </ td >\n          < td > Female </ td >\n     < td > No </ td >\n      < tr >\n            < td > Sun </ td >\n
  < td > Dinner </ td >\n        < td > 4 </ td >\n          </ tr >\n         < tr >\n          < td > 15.42 </ td >\n
  < td > 1.57 </ td >\n          < td > Male </ td >\n       < td > No </ td >\n      < tr >\n            < td > Sun </ td >\n
  < td > Dinner </ td >\n        < td > 2 </ td >\n          </ tr >\n         < tr >\n          < td > 18.43 </ td >\n
  < td > 3.00 </ td >\n          < td > Male </ td >\n       < td > No </ td >\n      < tr >\n            < td > Sun </ td >\n
  < td > Dinner </ td >\n        < td > 4 </ td >\n          </ tr >\n         < tr >\n          < td > 14.83 </ td >\n
  < td > 3.02 </ td >\n          < td > Female </ td >\n     < td > No </ td >\n      < tr >\n            < td > Sun </ td >\n
  < td > Dinner </ td >\n        < td > 2 </ td >\n          </ tr >\n         < tr >\n          < td > 21.58 </ td >\n
  < td > 3.92 </ td >\n          < td > Male </ td >\n       < td > No </ td >\n      < tr >\n            < td > Sun </ td >\n
  < td > Dinner </ td >\n        < td > 2 </ td >\n          </ tr >\n         < tr >\n          < td > 10.33 </ td >\n
  < td > 1.67 </ td >\n          < td > Female </ td >\n     < td > No </ td >\n      < tr >\n            < td > Sun </ td >\n
  < td > Dinner </ td >\n        < td > 3 </ td >\n          </ tr >\n         < tr >\n          < td > 16.29 </ td >\n
  < td > 3.71 </ td >\n          < td > Male </ td >\n       < td > No </ td >\n      < tr >\n            < td > Sun </ td >\n
  < td > Dinner </ td >\n        < td > 3 </ td >\n          </ tr >\n         < tr >\n          < td > 16.97 </ td >\n
  < td > 3.50 </ td >\n          < td > Female </ td >\n     < td > No </ td >\n      < tr >\n            < td > Sun </ td >\n
  < td > Dinner </ td >\n        < td > 3 </ td >\n          </ tr >\n         < tr >\n          < td > 20.65 </ td >\n
  < td > 3.35 </ td >\n          < td > Male </ td >\n       < td > No </ td >\n      < tr >\n            < td > Sat </ td >\n
  < td > Dinner </ td >\n        < td > 3 </ td >\n          </ tr >\n </ tbody >\n </ table ></ div >'
>>>
```

由上可见,该代码获取的数据相对简单,但与前面采用静态表的示例不同,网页中的数据不仅仅是表数据。下一步是将其传递到 BeautifulSoup 以实现 HTML 的解析,示例代码及结果如下[1]:

```
>>> soup = BeautifulSoup(r.text, "html.parser")
>>> soup.find('table', id = 'tips')
< table border = "6" class = "dataframe" id = "tips">
< thead >
< tr style = "text-align: center;">
```

---

[1] find()函数返回的数据太多,在……处省略了部分数据,原书未标出。

```html
<th>total_bill</th>
<th>tip</th>
<th>sex</th>
<th>smoker</th>
<th>day</th>
<th>time</th>
<th>size</th>
</tr>
</thead>
<tbody>
<tr>
<td>16.99</td>
<td>1.01</td>
<td>Female</td>
<td>No</td>
<td>Sun</td>
<td>Dinner</td>
<td>2</td>
</tr>
<tr>
<td>10.34</td>
<td>1.66</td>
<td>Male</td>
<td>No</td>
<td>Sun</td>
<td>Dinner</td>
<td>3</td>
</tr>
……
<tr>
<td>16.97</td>
<td>3.50</td>
<td>Female</td>
<td>No</td>
<td>Sun</td>
<td>Dinner</td>
<td>3</td>
</tr>
<tr>
<td>20.65</td>
<td>3.35</td>
<td>Male</td>
<td>No</td>
<td>Sat</td>
<td>Dinner</td>
<td>3</td>
</tr>
</tbody>
</table>
>>>
```

在使用表 id 时，我们可以直接转到 HTML 中的表，然后可以像以前一样访问其中所有的行。注意，我们只显示了该数据的一个子集，即前 20 行数据。现在，如果想从 HTML 解析数据，可以使用类似于在虚拟数据（dummy data）上使用的方法，示例代码如下：

```python
>>> table = soup.find('table', id='tips')
>>> table_rows = table.find_all("tr")
>>> table_rows[0:3]
[<tr style="text-align: center;">
<th>total_bill</th>
<th>tip</th>
<th>sex</th>
<th>smoker</th>
<th>day</th>
<th>time</th>
<th>size</th>
</tr>, <tr>
<td>16.99</td>
<td>1.01</td>
<td>Female</td>
<td>No</td>
<td>Sun</td>
<td>Dinner</td>
<td>2</td>
</tr>, <tr>
<td>10.34</td>
<td>1.66</td>
<td>Male</td>
<td>No</td>
<td>Sun</td>
<td>Dinner</td>
<td>3</td>
</tr>]
>>> headers = []
>>> content = []
>>> for tr in table_rows:
...     header_tags = tr.find_all("th")
...     if len(header_tags) > 0:
...         for ht in header_tags:
...             headers.append(ht.text)
...     else:
...         row = []
...         row_tags = tr.find_all("td")
...         for rt in row_tags:
...             row.append(rt.text)
...         content.append(row)
...
>>> headers
```

```
['total_bill', 'tip', 'sex', 'smoker', 'day', 'time', 'size']
>>> content
[['16.99', '1.01', 'Female', 'No', 'Sun', 'Dinner', '2'], ['10.34', '1.66', 'Male', 'No', 'Sun',
'Dinner', '3'], ['21.01', '3.50', 'Male', 'No', 'Sun', 'Dinner', '3'], ['23.68', '3.31', 'Male', 'No',
'Sun', 'Dinner', '2'], ['24.59', '3.61', 'Female', 'No', 'Sun', 'Dinner', '4'], ['25.29', '4.71',
'Male', 'No', 'Sun', 'Dinner', '4'], ['8.77', '2.00', 'Male', 'No', 'Sun', 'Dinner', '2'], ['26.88',
'3.12', 'Male', 'No', 'Sun', 'Dinner', '4'], ['15.04', '1.96', 'Male', 'No', 'Sun', 'Dinner', '2'],
['14.78', '3.23', 'Male', 'No', 'Sun', 'Dinner', '2'], ['10.27', '1.71', 'Male', 'No', 'Sun',
'Dinner', '2'], ['35.26', '5.00', 'Female', 'No', 'Sun', 'Dinner', '4'], ['15.42', '1.57', 'Male',
'No', 'Sun', 'Dinner', '2'], ['18.43', '3.00', 'Male', 'No', 'Sun', 'Dinner', '4'], ['14.83', '3.02',
'Female', 'No', 'Sun', 'Dinner', '2'], ['21.58', '3.92', 'Male', 'No', 'Sun', 'Dinner', '2'], ['10.33',
'1.67', 'Female', 'No', 'Sun', 'Dinner', '3'], ['16.29', '3.71', 'Male', 'No', 'Sun', 'Dinner', '3'],
['16.97', '3.50', 'Female', 'No', 'Sun', 'Dinner', '3'], ['20.65', '3.35', 'Male', 'No', 'Sat',
'Dinner', '3']]
>>>
```

正如上面所看到的,我们现在已经从 HTML 中提取了数据,并将其放入 headers 和 content 两个单独的列表中。进而,可以简单地利用本书前面介绍的方法将其放回 DataFrame 中。

```
>>> import pandas as pd
>>> data = pd.DataFrame(content)
>>> data
        0     1       2       3    4       5  6
0   16.99  1.01  Female      No  Sun  Dinner  2
1   10.34  1.66    Male      No  Sun  Dinner  3
2   21.01  3.50    Male      No  Sun  Dinner  3
3   23.68  3.31    Male      No  Sun  Dinner  2
4   24.59  3.61  Female      No  Sun  Dinner  4
5   25.29  4.71    Male      No  Sun  Dinner  4
6    8.77  2.00    Male      No  Sun  Dinner  2
7   26.88  3.12    Male      No  Sun  Dinner  4
8   15.04  1.96    Male      No  Sun  Dinner  2
9   14.78  3.23    Male      No  Sun  Dinner  2
10  10.27  1.71    Male      No  Sun  Dinner  2
11  35.26  5.00  Female      No  Sun  Dinner  4
12  15.42  1.57    Male      No  Sun  Dinner  2
13  18.43  3.00    Male      No  Sun  Dinner  4
14  14.83  3.02  Female      No  Sun  Dinner  2
15  21.58  3.92    Male      No  Sun  Dinner  2
16  10.33  1.67  Female      No  Sun  Dinner  3
17  16.29  3.71    Male      No  Sun  Dinner  3
18  16.97  3.50  Female      No  Sun  Dinner  3
19  20.65  3.35    Male      No  Sat  Dinner  3
>>> data.columns = headers
>>> data
   total_bill   tip     sex  smoker  day    time  size
0       16.99  1.01  Female      No  Sun  Dinner     2
1       10.34  1.66    Male      No  Sun  Dinner     3
```

2	21.01	3.50	Male	No	Sun	Dinner	3
3	23.68	3.31	Male	No	Sun	Dinner	2
4	24.59	3.61	Female	No	Sun	Dinner	4
5	25.29	4.71	Male	No	Sun	Dinner	4
6	8.77	2.00	Male	No	Sun	Dinner	2
7	26.88	3.12	Male	No	Sun	Dinner	4
8	15.04	1.96	Male	No	Sun	Dinner	2
9	14.78	3.23	Male	No	Sun	Dinner	2
10	10.27	1.71	Male	No	Sun	Dinner	2
11	35.26	5.00	Female	No	Sun	Dinner	4
12	15.42	1.57	Male	No	Sun	Dinner	2
13	18.43	3.00	Male	No	Sun	Dinner	4
14	14.83	3.02	Female	No	Sun	Dinner	2
15	21.58	3.92	Male	No	Sun	Dinner	2
16	10.33	1.67	Female	No	Sun	Dinner	3
17	16.29	3.71	Male	No	Sun	Dinner	3
18	16.97	3.50	Female	No	Sun	Dinner	3
19	20.65	3.35	Male	No	Sat	Dinner	3

\>>>

至此，我们已经完成了一个完整的网页爬取的过程：使用 DataFrame 在网站中填充一个表，然后抓取数据并将其转换回 DataFrame。

本章小结

本章介绍的内容很多，从 HTML 简介到 HTML 的解析，再到构建自己的网站，并从中抓取信息。本章所举示例主要集中于表数据，但可以应用于在 HTML 中找到的任何数据。在进行网络抓取时，Python 是一种十分流行且功能强大的工具，可以在网络中进行交互并获取数据。

CHAPTER 20

第 20 章 总　　结

  本书讲解了 Python 的相关知识，涵盖了 Python 所有的基础知识，给出了一些复杂的示例。然而，限于篇幅，还有很多内容没有涉及，需要读者自主学习。Python 具有众多的包，这些包的内容一直在持续改进，学习 Python 就要跟上语言的发展趋势，这一点十分重要。从开发的角度来看，我们采用了在 shell 中进行编码的方式，保持了脚本编写的简单性。然而，Python 绝不仅仅是一种探索性语言，Python 可以很好地应用在产品环境中。由于 Python 可以很好地应用于云计算解决方案、Web 应用程序开发、机器学习等，因此被许多大型科技公司采用。

  本书将为您打开一扇大门，让您开启自己的 Python 之旅，在庞大的 Python 社区中，找到自己想要学习的内容，开拓学习新的技术知识。祝您好运！